Biography

Pat Perry, MCIEH, MIOSH, FRSH, MIIRM, qualified as an Environmental Health Officer in 1978 and spent the first years of her career in local government enforcing environmental health laws, in particular health and safety law, which became her passion. She has extensive knowledge of her subject and has served on various working parties on both health and safety and food safety. Pat contributes regularly to professional journals, e.g. *Facilities Business,* and has been commissioned by Thomas Telford Publishing to write a series of health and safety books.

After a period in the private sector, Pat set up her own environmental health consultancy, Perry Scott Nash Associates Ltd, in the latter part of 1988, and fulfilled her vision of a 'one-stop shop' for the provision of consultancy services to the commercial and retail sectors.

The consultancy has grown considerably over the years and provides consultancy advice to a wide range of clients in a variety of market sectors. Leisure and retail have become the consultancy's major expertise and the role of planning supervisor and environmental health consultant is provided on projects ranging from a few hundred thousand pounds to many millions, e.g. new public house developments and major department store refits and refurbishments.

Perry Scott Nash Associates Ltd have strong links to the enforcing agencies; consultants having come mostly from similar backgrounds and approach projects and all the issues and concerns associated with legal compliance with pragmatism and commercial understanding.

Should you wish to contact Pat Perry about any issue in this book, or to enquire further about the consultancy services offered by Perry Scott Nash Associates Ltd, please contact us direct at:

Perry Scott Nash Associates Ltd
Perry Scott Nash House
Primett Road
Stevenage
Herts
SG1 3EE

Alternatively phone, fax or email on:

Tel: 01438 745771
Fax: 01438 745772
Email: p.perry@perryscottnash.co.uk

We would also recommend that you visit our website at:
www.perryscottnash.co.uk

Fire safety questions and answers: a practical approach

Pat Perry

Published by Thomas Telford Publishing, Thomas Telford Ltd, 1 Heron Quay, London E14 4JD.
URL: http://www.thomastelford.com

Distributors for Thomas Telford books are
USA: ASCE Press, 1801 Alexander Bell Drive, Reston, VA 20191-4400, USA
Japan: Maruzen Co. Ltd, Book Department, 3–10 Nihonbashi 2-chome, Chuo-ku, Tokyo 103
Australia: DA Books and Journals, 648 Whitehorse Road, Mitcham 3132, Victoria

First published 2003

Also in this series from Thomas Telford Books
Construction safety: questions and answers. Pat Perry. ISBN 0 7277 3233 1
Health and safety: questions and answers. Pat Perry. ISBN 07277 3240 4
Risk assessment: questions and answers. Pat Perry. ISBN 0 7277 3238 2
CDM questions and answers: a practical approach 2nd edition. Pat Perry. ISBN 0 7277 3107 6

A catalogue record for this book is available from the British Library

ISBN: 0 7277 3239 0

Any safety sign or symbol used in this book is for illustrative purposes only and does not necessarily imply that the sign or symbol used meets any legal requirements or good practice guides. Before producing any sign or symbol, the reader is recommended to check with the relevant British Standard or the Health and Safety (Safety Signs and Signals) Regulations 1996.

Throughout the book the personal pronouns 'he', 'his', etc. are used when referring to 'the Client', 'the Designer', 'the Planning Supervisor', etc., for reasons of readability. Clearly, it is quite possible these hypothetical characters may be female in 'real-life' situations, so readers should consider these pronouns to be grammatically neuter in gender, rather than masculine.

Typeset by Alex Lazarou, Surbiton, Surrey
Printed and bound in Great Britain by MPG Books, Bodmin, Cornwall

Acknowledgements

My sincere thanks go to Maureen for her never ending support and encouragement and to Janine and the Business Support team at Perry Scott Nash Associates Ltd for typing all the handwritten manuscripts with such patience and efficiency.

Author's note

Many of the publications referenced in this book are available for download on a number of websites, e.g.:

- www.hse.gov.uk/pubns/index.htm
- www.hsebooks.co.uk

Also, guidance on the availability of books is available from the HSE Info Line on 0541 545 500.

Contents

1

Introduction to the principles of fire

What is fire?

A standard definition of fire is:

> a process of combustion characterised by the emission of heat accompanied by smoke or flame.

Mostly, when the term fire is referred to, we mean the uncontrolled development of fire not a controlled fire such as happens in furnaces, etc.

Fire is often unwanted, unexpected, disastrous and costly, both in terms of human life and business costs.

What causes a fire?

There is an interrelationship between heat, fuel and oxygen. This is known as the *fire triangle*.

Fire is the result of a chemical reaction between a fuel and oxygen. Fire cannot occur if one of the key components is missing, i.e. if the heat, the fuel or oxygen is eliminated.

This is the principle under which fires are fought:

- drench with water thereby removing the heat
- smother with inert material (e.g. sand), thereby depriving the fire of oxygen
- remove the fuel, i.e. separate burning material from other materials.

The triangle of fire

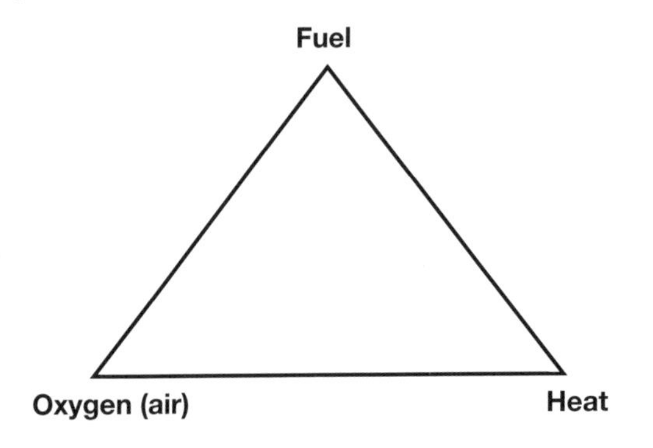

Is it the actual material that burns?

In many cases it is the body of the material which burns (e.g. wood or clothing), but in many other instances it is also the gaseous vapours which are given off during the burning process which re-ignite and cause the fire to develop.

How do materials catch fire?

Heat or thermal energy is transferred from hot materials to cooler ones by radiation, conduction and convection.

Thermal radiation, like light, can travel across empty space and the amount which is transferred from one place to another follows the same rules as light.

Conduction is the direct transfer of heat energy to an adjoining substance, e.g. placing something on the radiator to dry or 'heat up'.

Convection is the transfer of heat by movement caused by changes in air density.

The most important means of fire spread in a building are convection and radiation.

How quickly does fire spread?

In nine cases out of ten, very quickly.

Fire development is often described as exponential — in simple terms, fire tends to double its size at regular intervals — sometimes as quickly as within minutes.

Fires in buildings develop differently to fires in the outdoors (e.g. bonfires), and the speed at which they spread takes most people completely by surprise.

It is said that the average domestic sitting room with usual equipment of TV and video, sofas, etc. will be totally alight in about four minutes.

Often, members of the public fail to comprehend the speed at which fires develop and do not believe that they have to evacuate or do anything in relation to their safety.

The major fires in Woolworths, Manchester in 1979, the Dublin Disco in 1981 and the Bradford fire in 1985 are all examples of a rapidly spreading fire and people delaying in taking it seriously and evacuating the building.

Why is smoke such a hazard?

Smoke is defined as:

> the product of combustion, consisting of fine particles of carbon carried by hot gases and air.

Smoke spreads easily around a room or building and, scientifically, it has been proven that smoke moves faster than people can walk.

Smoke contains particulate matter and this makes the 'flame' opaque — so that you cannot see through it. Therefore it is a hazard because it cuts down visibility.

Smoke also contains the 'products of combustion', depending on what materials are burning, and contains poisonous gases and toxic fumes.

Common gases — the products of combustion — are carbon monoxide and hydrogen cyanide.

Carbon monoxide, an odourless, colourless gas, is lethal and kills within minutes at high concentrations. Most people caught in a fire die from breathing in carbon monoxide gas.

Before inhalation of the smoke becomes lethal it can cause severe respiratory irritation — breathing becomes difficult.

Carbon monoxide, in non-lethal concentrations, will affect a person's ability to concentrate.

Couple lack of concentration with visual difficulty and people quickly become disorientated in a smoke-filled environment.

This is one of the reasons why fire exit and emergency signage must be clear and properly lit.

People are reluctant to walk through smoke even though it would lead them to safety. People therefore become trapped, often by their own inability to determine the greater danger, and as a result suffer injury or death.

What does the term 'flashover' mean?

Flashover is defined as:

> a sudden transition to a state of total surface involvement in a fire of combustible materials within a compartment.

After the first item in an enclosure is ignited, hot gases rise vertically in a narrow plume into which air is pulled from the natural

Case study

The King's Cross Fire

The Fire Brigade was called at 19:36 to a fire on an escalator at King's Cross Underground Station. The fire seemed controllable and in one location. Passengers were evacuated up escalators via the ticketing hall. Tubes were still stopping and passengers were still embarking and disembarking.

At 19:45 the fire suddenly erupted up into the ticketing hall. It was described as 'flashing, with searing heat and thick smoke through the ticket hall'. 'Within seconds, the area was in total darkness and the conditions had become unbearable'. Escalators carried passengers up into the ticket hall and certain death.

Thirty-one people died as a result of the King's Cross fire and many more suffered catastrophic burns and other injuries.

The cause of the fire was established to be discarded smoking material which fell through a wooden escalator onto grease and debris which was extremely combustible.

ventilation in the room, so increasing the volume of the smoke and gases.

As smoke reaches the ceiling it spreads out in all directions and begins to form a thickening layer below the ceiling.

As the fire grows, the flames reach the ceiling and are deflected horizontally across the surface of the ceiling. Radiation is so strong that flame will spread over all combustible surfaces, and items over a relatively large area will reach temperatures at which they spontaneously burst into flames.

The fire is contained in the compartment (e.g. office or room), and when everything is alight this is known as 'flashover'.

At flashover, virtually all items in the room are alight because the room temperature has risen to in excess of 1000ºC and the survival of any occupants of the room is impossible.

What is flame spread?

Flame spread is continuous ignition. Flames spread over a solid surface as the surface in advance of the flame becomes heated by radiation, conduction or convected heat transfer, causing gaseous fuel to be released, so extending the burning zone.

The speed of advance of flames can be increased by air movement, e.g. a flame driven sideways by the wind extends the area of heating thus causing more heat, gases and fumes.

Flame spread is much quicker when the surfaces' temperatures have been pre-heated and rapid flame spread can lead to flashover.

What is spontaneous combustion?

The *Encyclopaedia Britannica* defines spontaneous combustion as 'the outbreak of fire without application of heat from an external source'.

This combustion can occur when flammable matter like oily rags, damp hay, leaves or coal is stored in bulk. Spontaneous combustion (sometimes called spontaneous ignition) begins when a combustible object is heated to its ignition temperature by a slow oxidation process. Oxidation is a chemical reaction involving the oxygen in the air around us gradually raising the inside temperature of something (e.g. a pile of oily rags), to a point at which a fire starts.

The way in which combustible materials are stored has a lot to do with whether they will spontaneously combust or not. Aerosols and other flammable liquids or gases often carry instructions about being stored away from a heat source. This prevents temperatures rising to the point of ignition.

Substances with low ignition temperatures pose a greater risk than others.

Why do buildings often collapse in a fire?

At some stage in the development of fire, the walls, floor and ceilings of a building are subject to heating, mostly in the form of radiation, at temperatures of many hundreds, if not thousands, of degrees.

The capacity of a building to withstand this heat and still support loadings and prevent the passage of heat and gas diminishes with the period of exposure to sustained heating.

Steel girders can buckle in the heat and lose their load bearing capacity.

Other materials burn to ashes and fail to support loads.

The structural integrity of a building during a fire is a critical part of fire protection and, without such fire protection, buildings would be extremely vulnerable to collapse.

What is meant by the classification of fires?

Fires are classified according to British Standard 4547, a harmonised European standard.

Fires are classified so that the suitability of different extinguishing agents can be determined. The classification depends on the actual type of combustible material involved in the fire.

Class A These fires involve solid material, usually of an organic nature, in which combustion normally takes place with the formation of embers.
Material types are wood, fibre, textile fabrics and paper.
Cooling by water is the most effective method of extinguishing this type of fire.

Class B These fires involve liquids or liquefied solids such as petrol, oil, greases or fats.
The most effective way of extinguishing this group of fires is by smothering the fire and depriving it of oxygen.
Types of extinguishing agent would be foam and powder.
A fire blanket can also be used in certain situations.

Class C These fires involve gases such as hydrogen, propane, butane and other liquefied petroleum gases.
Gaseous fires are extremely difficult to contain and the most effective way is to cut off the gas supply at source.
Any burning material set alight by the fire can be dealt with by the best medium.

Class D These fires involve metals and other substances such as magnesium, sodium, aluminium, potassium and calcium.
Special techniques for fire fighting are required.

What is an electrical fire?

Strictly speaking there is no such thing.

An electrical fire is often used to describe a fire in which electricity is present. The electricity may be supplying the source of heat (ignition) through electrical sparks or overheating and the fuel may be the electrical cabling and wiring.

A fire in an electrical appliance will fall into whatever category the burning material is.

Water should never be sprayed on to live electrics so water-based fire extinguishers are not suitable on live electrical fire but could be used once the electrical supply has been *switched off*.

The most suitable fire extinguishing medium to use on live electrical equipment is carbon dioxide. Halon gas can also be used but this is being phased out as it is an ozone depleting gas.

What are some of the major causes of fire?

Some of the most common causes of fire are:

- faulty electrical equipment
- smoking and smoking materials
- fat fryers and chip pans
- portable heaters
- refuse and rubbish accumulations
- flammable materials.

Arson is now a common cause of commercial fire and incidents are rising in respect of domestic premises.

Easy access to ignition sources and fuel make arson easy. The number of school fires due to arson is increasing. Preventing access to grounds, buildings, avoiding accumulations of combustible materials, etc. will help to prevent arson.

What are the statistics regarding fires in the UK?

The latest Fire Statistics Monitor was published in February 2003 and refers to the twelve-month period to March 2002.

In that twelve-month period, local authority fire brigades attended 1 027 800 fires and false alarms.

The total number of actual fires was 547 700 and, within that figure, 228 000 fires were associated with property, i.e. buildings or vehicles.

Fires in buildings represented 111 400, of that number over 68 000 were dwellings and over 43 000 were commercial buildings, schools and public buildings. Road vehicle fires accounted for 103 000 fires.

False alarms totalled 480 000.

The total number of deaths from fire rose during the twelve months compared to the previous year, from 559 up to 614. Of that number, 433 deaths occurred in dwellings.

Injuries rose slightly during the year to 17 200, with injuries associated with dwelling fires static at 13 900.

There were 53 300 accidental fires in dwellings during the twelve-month period.

Malicious fires — arson — continued to increase over the period to 124 800. One of the highest increases was in malicious road vehicle fires.

Grassland and outdoor refuse fires rose by 31% over the year to 305 000.

The statistics for 2001–2002 (i.e. latest available mid-2003) show that the incidence of fire is increasing and, despite increased fire safety awareness, the number of people dying from fires in buildings is rising.

Complacency in fire safety is therefore not an option.

2

Fire safety legislation

What is the main piece of legislation governing fire safety in buildings?

As of 2003, the current major legislation governing fire safety is the Fire Precautions Act 1971.

This Act was introduced during 1970 as a result of a major hotel fire in 1969.

The Fire Precautions Act 1971 set a new style for legislation in that it was one of the first 'enabling acts' and could be flexible in responding to fire safety requirements. Premises would be 'designated' by an order as required to comply with the Act.

What are the main provisions of the Fire Precautions Act 1971?

The main part of the Act is the requirement for a *Fire Certificate*. This is issued by the Fire Authority and is required when:

- 20 or more people are employed on the premises at any one time
- 10 or more people are employed elsewhere than the ground floor
- sleeping accommodation is provided for more than 6 people.

This means that if you have 20 or more staff at any one time you will need to have a Fire Certificate. Also, if you have people working in a first floor area and the number of employees is 10 or more, you will need a Fire Certificate. Basement and other floor levels also count.

The Fire Authority can stipulate that other conditions are included in the Fire Certificate but, generally, a Certificate covers:

- the use of the building coved by the Certificate
- the means of escape provided
- safe use at all times of the means of escape
- the means for fighting a fire
- the means for giving warning
- staff training and records
- maintenance of means of escape
- limits on occupancy for the premises.

If you have a Fire Certificate and fail to comply with it, you commit an offence and the premises could be closed down.

If you have not yet obtained a Fire Certificate but you fall into the category of needing one, then as long as you have *applied* to the Fire Authority you are not committing an offence. In these circumstances, the law requires that you manage fire safety.

In addition to the need to apply for Fire Certificates, the 1971 Act gives Fire Authority Inspectors powers to inspect premises.

As long as a Fire Certificate is in force for your premises, a fire officer has the right to inspect at all reasonable times in order to ensure that the conditions laid down in the Fire Certificate are being met.

The Fire Authority are particularly interested in finding any variations to a Fire Certificate's condition due to internal alterations, etc. Many internal alterations lead to a reduction in fire safety as means of escape, travel distances, etc. can be compromised.

The Act contains a provision which enables the Fire Authority to grant certain premises an exemption from the need to have a Fire Certificate. Exemptions are granted at the Authority's discretion and if they think that, having considered all the circumstances, the seriousness of the risk to persons in the premises from fire is low, they can issue an Exemption Notice in writing.

The Fire Authority can withdraw the exemption if circumstances change at any time, provided that they have given the building owners or occupiers notice of their intention to withdraw the exemption.

Where an exemption is granted, owners or occupiers must give early notification to the Fire Authority of any intended building works and must not commence those works until authorised to do so.

What are the powers of Fire Authority Inspectors?

Fire Authority Inspectors have many similar powers to HSE Inspectors and Local Authority EHOs.

Fire Authority Inspectors can:

- enter buildings and inspect all or part of them at all reasonable times
- make any inquiries as may be necessary in order to establish whether a Fire Certificate is needed or to identify the owner or occupier
- make enquiries as to the identity of any manager of the building
- make enquiries as to whether the requirements of the Fire Certificate are being followed
- inspect documents, especially the Fire Certificate
- request assistance in their duties from anyone who has responsibilities in respect of the building
- serve 24 hours' notice of intended entry on a domestic dwelling and thereafter demand entry for inspection.

If Fire Authority Inspectors find contraventions of the Fire Precautions Act 1971, in particular the requirements of the Fire Certificate, they can serve Improvement or Prohibition Notices on either the owner or occupier.

What is an Improvement Notice?

An Improvement Notice is a formal notice issued by the Fire Authority when they believe that the duties imposed by the Act are being contravened.

Improvement Notices are generally served when there has been a contravention in providing adequate means of escape from the building and also, or in addition, inadequate fire-fighting provision has been made.

An Improvement Notice can be served on all premises, including those that are not required to have a Fire Certificate. Even if a Fire Certificate is not required, building owners and occupiers have to take reasonable precautions to safeguard people from the consequences of fire.

An Improvement Notice must state what requirement has been contravened and must state what needs to be done to put it right. A time limit for compliance must be stated.

What is a Prohibition Notice?

A Prohibition Notice is a formal notice served by the Fire Authority on any premises when they consider that a dangerous situation exits in relation to fire safety.

The notice can prohibit or restrict the use of the whole building or only parts of it. It can prohibit a certain use or can stop excess numbers of people resorting to the premises.

A Prohibition Notice is instant and takes immediate effect. It is used to prevent imminent risk of personal injury or harm due to fire.

The Fire Officer needs to be of the opinion that there is, or will be, a risk to persons on the premises from fire which is so serious that the premises ought to be prohibited or restricted.

The Notice must contain information about the risk to persons' safety, specify the matters which give rise to the risk and stipulate the prohibited or restricted use of the building until the specified matters have been remedied.

A Prohibition Notice may include directions as to what steps need to be taken to remedy the specified matters.

What are the penalties for contravening a Prohibition or Improvement Notice?

Penalties are similar to those relating to health and safety, i.e. fines and/or imprisonment.

If the case of failure to comply comes before the magistrates' court, the fines for contravening either type of notice will be up to £20 000.

If the case is heard in the Crown Court, fines can be unlimited and a prison sentence of up to two years can be imposed in addition.

If serious consequences arose as a result of failure to comply with a Prohibition Notice, the penalty would be likely to be a huge fine and imprisonment.

Individuals of the 'body corporate' could face personal prosecution if they were shown to have contributed to the offence by connivance, consent or neglect.

What are the Fire Precautions (Workplace) Regulations 1997 (as amended in 1999)?

The Fire Precautions (Workplace) Regulations 1997 came into force in December 1997 as a result of an EC Directive which required the harmonisation of certain health and safety standards across Europe.

The Regulations were amended slightly in 1999 because the EC did not think that the UK had implemented the Directive effectively.

The Regulations broke with tradition in respect of fire safety law because they were self-regulatory and not 'prescriptive'. UK laws had for many years been based on enforcing authorities inspecting premises and telling owners/occupiers what to do by way of letter or formal notice. A period of time was given to comply and then the inspecting officer returned to see that all had been accomplished.

The 'Workplace' Regulations introduced the concept of risk assessment into Fire Legislation, following the principles laid down in earlier EC Directives and UK legislation known as the 'six pack'.

However, the 1997 Regulations did not apply risk assessment requirements to all businesses, contrary to the EC Directive. The amendments made in 1999 *did* apply risk assessment to all workplaces in which employees were engaged at work.

The requirement for a Risk Assessment under the Fire Regulations follows the requirement for general risk assessments as laid down in the Management of Health and Safety at Work Regulations 1999.

Every employer has to assess the risks to the health and safety of his employees while at work.

The employer must also assess the risks to the health and safety of those not at work but affected by his undertaking, i.e. business activity.

The Fire Precautions (Workplace) Regulations 1997 (amended) also require employers to:

- establish and implement procedures to be followed in the event of serious or imminent danger
- provide information, instruction and training to employees
- co-operate with other employers in respect of fire safety, especially in a multi-occupied building
- write down arrangements for the effective health and safety management of the protective and preventative measures needed to ensure safety.

The Regulations introduced what have commonly become known as 'Fire Risk Assessments'.

How do the Building Regulations affect fire safety?

The Building Regulations set down standards to which new and refurbished buildings must conform in terms of construction design and safety.

Fire safety requirements for new building works and renovations, alteration, refurbishments and extension to existing buildings are covered in Part B of the Regulations.

The Building Regulations themselves are goal-setting and are supported by 'approved documents' which set out some practical ways in which to meet the legal requirements.

The fire safety aspects of Part B cover:

- means of escape
- fire alarms
- fire spread
- access and facilities for the Fire Service.

The Building Regulations are enforced by the Building Control Department of the Local Authority and a failure to comply with the requirements could lead to fines and an enforcement notice.

Building Control Departments must consult with the Fire Authority when necessary.

If a building complies with Part B of the Building Regulations it will, in most cases, comply with Fire Safety Legislation. However, there may be operational matters which are outside the remit of Building Regulations but for which the Fire Authority may require additional controls or standards.

There are a number of premises which are subject to various licensing conditions, e.g. residential care homes, hospitals, boarding schools, licensed premises, petrol filling stations, etc. Should this be the case, the Building Control Department is obliged to consult with all interested bodies, including the Licensing Authority, regarding fire safety issues.

What proposals are there to change the Fire Safety Legislation?

A Consultation Paper was issued in 2002 by the Office of the Deputy Prime Minister outlining proposed changes to the legislative regime for fire safety. The responses to the discussion process are being considered and it is anticipated that in 2004 there will be a new Regulatory Reform (Fire Safety) Order which will set out revised responsibilities for fire safety.

The key points of the Consultation Paper are detailed below.

- The Fire Precautions Act 1971 and the Fire Safety (Workplace) Regulations 1997 (as amended) will be replaced by one single piece of legislation called the Regulatory Reform (Fire Safety) order.
- The new legislation will be enforced mainly by fire authorities.
- Occupiers of premises currently designated as requiring a Fire Certificate under the Fire Precautions Act 1971 will no longer have to apply for one. None will be issued.
- Fire precautions will not be reduced overall but will be replaced by equal requirements under a new regime.
- Fire authorities will be given a new duty to promote community fire safety.
- The new proposals apply to England and Wales only. Scotland will remain as is until the Scottish Parliament amends its legislation.
- The new Fire Safety Order will be based around a new general duty of fire safety care, with specific requirements that will need to be met to comply with that duty.
- Fire safety will be based on a risk-based approach. Risk assessment will need to be carried out for all workplaces.
- With the abolition of Fire Certificates, the risk-based approach will apply to non-employees.
- The existing power for the Secretary of State to make regulations to address key fire safety issues will be retained.

- A 'responsible person' will need to be designated to comply with fire safety legislation.
- Empty buildings would be brought within the order and the owner would be responsible for fire safety.
- People employed to undertake duties that have a bearing on the safety of a building must be competent to do the job. The responsible person must ensure that people, including contractors, are competent.
- Legal action would be taken against a contractor if they had been negligent in their duties or if they had failed to complete their duties or misrepresented themselves in any way. Action could also be taken against the responsible person for failing to appoint a competent contractor.
- The new fire safety regime would cover all workplaces and places to which the public has access. Domestic premises would be excluded.
- Voluntary workers operating in premises will be brought into the new regime. They will have to be protected for fire safety.
- The responsible person will have a general duty to ensure the fire safety of others.
- Preventative and protective fire safety measures must be implemented by the responsible person.
- Fire safety arrangements must be made by the responsible person.
- Dangerous substances will have special arrangements and requirements.
- Provision must be made for fire fighting and fire detection.
- Emergency exit routes and exits must be designated and maintained.
- Premises and any fire-fighting equipment must be maintained by the responsible person.
- The responsible person must appoint a competent person to assist him in his duties of fire safety.
- Emergency procedures must be put in place for serious and imminent danger, and for danger areas.

- Responsible persons shall ensure that adequate information is available to those who will need it.
- Where there are two or more employers within a building there shall be co-operation and co-ordination to comply with the fire safety regime.
- Information must be given to visiting employees, those working away or in host employers' premises.
- Employees must be given general fire safety training and must be given specific training on risks, changes in responsibilities, etc.
- Employees will have general duties to take reasonable care of themselves and others for fire safety.

References

Fire Precautions Act 1971.

Fire Precautions (Workplace) Regulations 1997 (amended).

Management of Health and Safety at Work Regulations 1999.

Health and Safety at Work Etc. Act 1974.

3

Fire Risk Assessments

What is the legal requirement for a Fire Risk Assessment?

The Fire Precautions (Workplace) Regulations 1997, as amended in 1999, set down the requirements for employers to carry out a Fire Risk Assessment for all workplaces.

Any premises in which persons are employed must be subject to a Fire Risk Assessment.

The principles of risk assessment to be followed are those listed in the Management of Health and Safety at Work Regulations 1999.

Where there are five or more employees, the significant findings of the Risk Assessment must be recorded in writing.

Are premises with a Fire Certificate exempt from the need to carry out a Fire Risk Assessment?

No. Premises with a Fire Certificate were originally exempt from completing Fire Risk Assessments under the Fire Precautions (Workplace) Regulations 1997 but the European Commission declared that the UK had not implemented the European Directive correctly and required amendments to be made.

Amendments to the original 1997 Regulations were made in 1999 and a revised legal requirement was implemented which covered all workplaces. The 1997 Regulations are now read with the amendment Regulations.

The Management of Health and Safety at Work Regulations were also amended in 1999 and the requirement to complete a Fire Risk Assessment was inserted into Regulation 3 on risk assessments.

What actually is a Fire Risk Assessment?

A Fire Risk Assessment is, in effect, an audit of your workplace and work activities in order to establish how likely a fire is to start, where it would be, how severe it might be, who it would affect and how people would get out of the building in an emergency.

A Fire Risk Assessment should be concerned with *life safety* and not with matters which are really fire engineering matters.

A Fire Risk Assessment is a structured way of looking at the hazards and risks associated with fire and the products of fire, e.g. smoke.

Like all Risk Assessments, a Fire Risk Assessment follows *five key steps*, namely:

Step 1: Identify the hazards
Step 2: Identify the people and the location of people at significant risk from a fire
Step 3: Evaluate the risks
Step 4: Record findings and actions taken
Step 5: Keep assessment under review.

So, a Fire Risk Assessment is a record that shows you have assessed the likelihood of a fire occurring in your workplace, identified who could be harmed and how and decided on what steps you need to take to reduce the likelihood of a fire (and therefore its harmful consequences) occurring. You have recorded all these findings regarding your undertaking into a particular format, called a Risk Assessment.

Definitions

Risk Assessment — the overall process of estimating the magnitude of risk and deciding whether or not the risk is tolerable or acceptable.

Risk — the combination of the likelihood and consequence of a specified hazardous event occurring.

Hazard — a source or a situation with a potential to harm in terms of human injury or ill-health, damage to property, damage to the environment or a combination of these factors.

Hazard identification — the process of recognising that a hazard exists and defining its characteristics.

What are the *five steps* to a Fire Risk Assessment?

Step 1. Identify the hazards

Sources of ignition

You can identify the sources of ignition in your premises by looking for possible sources of heat that could get hot enough to ignite the material in the vicinity.

Such sources of heat/ignition could be:

- smokers' materials
- naked flames, e.g. candles, fires, blow lamps, etc.
- electrical, gas or oil-fired heaters
- Hot Work processes, e.g. welding or gas cutting
- cooking, especially frying

- faulty or misused electrical appliances including plugs and extension leads
- lighting equipment, especially halogen lamps
- hot surfaces and obstructions of ventilation grills, e.g. radiators
- poorly maintained equipment that causes friction or sparks
- static electricity
- arson.

Look out for evidence that any items have suffered scorching or overheating, e.g. blackened plugs and sockets, burn marks, cigarette burns, scorch marks, etc.

Check each area of the premises systematically:

- customer areas, public areas and reception
- work areas and offices
- staff kitchen and staff rooms
- store rooms and cleaners' stores
- plant rooms and motor rooms
- refuse areas
- external areas.

Sources of fuel

Generally, anything that burns is fuel for a fire. Fuel can also be invisible in the form of vapours, fumes, etc. given off from other less flammable materials.

Look for anything in the premises that is in sufficient quantity to burn reasonably easily, or to cause a fire to spread to more easily combustible fuels.

Fuels to look out for are:

- wood, paper or cardboard
- flammable chemicals, e.g. cleaning materials
- flammable liquids, e.g. cleaning substances, liquid petroleum gas
- flammable liquids and solvents, e.g. white spirit, petrol, methylated spirit

- paints, varnishes, thinners, etc.
- furniture, fixtures and fittings
- textiles
- ceiling tiles and polystyrene products
- waste materials and general rubbish
- gases.

Consider also the construction of the premises — are there any materials used which would burn more easily than other types. Hardboard, chipboard and blockboard burn more easily than plasterboard.

Identifying sources of oxygen

Oxygen is all round us in the air that we breathe. Sometimes, other sources of oxygen are present that accelerate the speed at which a fire ignites, e.g. oxygen cylinders for welding.

The more turbulent the air, the more likely the spread of fire will be, e.g. opening doors brings a 'whoosh' of air into a room and the fire is fanned and intensifies. Mechanical ventilation also moves air around in greater volumes and more quickly.

Do not forget that while ventilation systems move oxygen around at greater volumes, they will also transport smoke and toxic fumes around the building.

Step 2. Identify who could be harmed

You need to identify who will be at risk from a fire and where they will be when a fire starts. The law requires you to ensure the safety of your staff and others, e.g. customers. Would anyone be affected by a fire in an area that is isolated? Could everyone respond to an alarm, or evacuate?

Will you have people with disabilities in the premises, e.g. wheelchair users, visually or hearing impaired? Will they be at any greater risk of being harmed by a fire than other people?

Will contractors working in plant rooms, on the roof, etc. be adversely affected by a fire? Could they be trapped or fail to hear alarms?

Who might be affected by smoke travelling through the building? Smoke often contains toxic fumes.

Step 3. Evaluate the risks arising from the hazards

What will happen if there is a fire? Does it matter whether it is a minor or major fire? Remember that small fires can grow rapidly into infernos.

A fire is often likely to start because:

- people are careless with cigarettes and matches
- people purposely set light to things
- cooking canopies catch fire due to grease build-up
- people put combustible material near flames/ignition sources
- equipment is faulty because it is not maintained
- electrical sockets are overloaded.

Will people die in a fire from:

- flames
- heat
- smoke
- toxic fumes?

Will people be trapped in the building?

Will people know that there is a fire and will they be able to get out?

Step 3 of the Risk Assessment is about looking at what *control measures* you have in place to help control the risk or reduce the risk of harm from a fire.

Remember — fire safety is about *life safety*. Get people out fast and protect their lives. Property is always replaceable.

You will need to record on your Fire Risk Assessment the fire precautions you have in place, i.e.:

- What emergency exits do you have and are they adequate and in the correct place?
- Are they easily identified and unobstructed?
- Is there fire-fighting equipment?
- How is the fire alarm raised?
- Where do people go when they leave the building — an assembly point?
- Are the signs for fire safety adequate?
- Who will check the building and take charge of an incident, i.e. do you have a Fire Warden appointed?
- Are fire doors kept closed?
- Are ignition sources controlled and fuel sources managed?
- Do you have procedures to manage contractors? (Remember that Windsor Castle went up in flames because a contractor used a blow torch near the curtains!)

Taking all your fire safety precautions for the premises into consideration, is there anything more that you need to do?

Are staff trained in what to do in an emergency? Can they use fire extinguishers? Do you have fire drills? Is equipment serviced and checked, e.g. emergency lights, fire alarm bells, etc.

Step 4. Record findings and action taken

Complete a Fire Risk Assessment form and keep it safe.

Make sure that you share the information with staff.

If contractors come to site, make sure that you discuss *their* fire safety plans with them and that you tell them what your fire precaution procedures are.

Step 5. Keep Assessment under review

A Fire Risk Assessment needs to be reviewed regularly — about every six months or so and whenever something has changed, including layout, new employees, new procedures, new legislation, increased stock, etc.

Is there any guidance on assessing the risk rating of premises in respect of fire safety?

When completing Fire Risk Assessments it is sensible to categorise *residual risk* for the premises into a risk rating category — normally referred to as high, medium or low.

In terms of fire risk rating it is usual to refer to medium risk as 'normal'.

The Government's publication *Fire safety — an employer's guide* gives some guidance on how to fire risk rate premises.

High risk premises

- Any premises where highly flammable or explosive substances are stored or used (other than in very small quantities).
- Any premises where the structural elements present are unsatisfactory in respect of fire safety:
  - lack of fire-resisting separation
  - vertical or horizontal openings through which fire, heat and smoke can spread
  - long and complex escape routes created by extensive sub-division of floors by partitions, etc.
  - complex escape routes created by the positioning of shop unit displays, machinery, etc.
  - large areas of smoke- or flame-producing furnishings and surface materials especially on walls and ceilings.
- Permanent or temporary work activities which have the potential for fires to start and spread, e.g.:
  - workshops using highly flammable materials and substances

- paint spraying
- activities using naked flames, e.g. blow torches and welding
- large kitchens in work canteens and restaurants
- refuse chambers and waste disposal areas
- areas containing foam or foam plastic upholstery and furniture.

- Where there is significant risk to life in case of fire:
  - sleeping accommodation provided for staff, guests, visitors, etc. in significant numbers
  - treatment or care where occupants have to rely on others to help them
  - high proportions of elderly or infirm
  - large numbers of people with disabilities
  - people working in remote areas, e.g. plant rooms, roof areas, etc.
  - large numbers of people resorting to the premises relative to its size, e.g. sales at retail shops
  - large numbers of people resorting to the premises where the number of people available to assist is limited, e.g. entertainment events, banquets, etc.

Normal risk premises

- Where an outbreak of fire is likely to remain contained to localised areas or is likely to spread only slowly, allowing people to escape to a place of safety.
- Where the number of people in the premises is small and they are likely to escape via well-defined means of escape to a place of safety without assistance.
- Where the premises have an automatic warning system or an effective automatic fire-fighting, fire-extinguishing or fire-suppression system which may reduce the risk categorisation from high.

Low risk premises

Where there is minimal risk to peoples' lives and where the risk of fire occurring is low or the potential for fire, heat or smoke spreading is negligible.

What type of people do I need to worry about when I carry out my Risk Assessment?

Employers must consider the following people as being at risk in the event of a fire:

- employees
- employees whose mobility, sight or hearing might be impaired
- employees with learning difficulties or mental illness
- other persons in the premises if the premises are multi-occupied
- anyone occupying remote areas of the premises
- visitors and members of the public
- anyone who may sleep on the premises.

Does a Fire Risk Assessment have to consider members of the public?

A Fire Risk Assessment must be carried out by an employer and must consider the risks to the safety of *employees*.

However, under the general provisions of the Management of Health and Safety at Work Regulations 1999, all persons who may be affected by the employer's business or undertaking must be considered in the Risk Assessment.

Case study

Fire hazards in licensed premises

Fires in public houses are quite common and usually occur in the kitchen or customer area where the risks of fire are greatest.

What do I need to look for?

Kitchen

- Storage of flammable materials, e.g. cardboard packaging near to a heat source.
- Grease build-up on grills, filters, cooking equipment as it can easily ignite and flames can spread rapidly.
- Grease and dirt build-up on canopies and within ductwork.
- Overheating deep fat fryers — oil left burning for long periods of time.
- Flammable materials next to cooker, grill, griddle or hob tops.
- Overloaded sockets.
- Poorly repaired plugs.
- Use of wrong fuses in plugs.
- Poor use of extension leads, wrong fuses or overloaded sockets.
- Poorly maintained equipment.
- Misuse of microwaves and use of combustible material as packaging, cardboard dishes, etc.
- Storage of flammable aerosols and cleaning fluids near to heat sources.
- Use of blowtorches for 'burnishing' toppings, glazing, etc.

- Generation of flammable fumes and vapours from aerosols, etc. used elsewhere but where the fumes drift and ignite near a heat source.
- Use of LPG in the kitchen.
- Faulty gas appliances — usually associated with explosions.
- Arson.

Bar servery

- Overloaded sockets.
- Disposal of cigarette debris, ashtrays, etc. into rubbish bins or waste bins which are combustible and which contain combustible material, e.g. packaging.
- Electric faults on equipment, e.g. glass washers and fridges.
- Spread of fire to the bar via python runs and other voids which communicate with the other areas of the pub.
- Overheating equipment due to vent grill obstructions.

Customer area

- Smoking materials dropped onto seating areas, carpeting, etc.
- Poorly extinguished cigarettes.
- Disposal of ashtrays into waste receptacles containing combustible materials.
- Overloaded electrical sockets.
- Furniture, etc. too close to real fires or gas fires.
- Light fittings with the wrong wattage bulbs, etc.
- Light fittings too near to combustible objects where heat transference can cause combustion.
- Arson.

Cellar

- Overloaded electrical sockets.
- Use of flammable cleaning fluids near electrical ignition sources.
- Storage of flammable substances near heat sources.
- Discarded smoking materials, i.e. hastily discarded cigarettes, matches, etc.
- Storage of combustible materials near a heat source.
- Blocking up equipment ventilation grills, e.g. ice machines, causing the equipment to overheat and spontaneously combust.
- Arson.

Plant rooms

- Poorly maintained equipment.
- Storage of combustible materials and substances near to heat sources.
- Electrical faults.
- Escape of combustible fumes, gases, etc.
- Overloaded electrical sockets and extension leads.
- Electrical arcing.
- Use of flammable cleaning chemicals.
- Grease and dirt build-up within equipment which is heat generating.

Staff areas

- Many of the incidents already listed can apply to staff areas.
- Faulty washing machine and tumble drier equipment.
- Poor ventilation to electrical equipment.
- Use of portable heating equipment.

Top tips

- Make regular checks to identify fire hazards.
- Look out for anything unusual — blackening of plugs, sockets, etc.
- Do not overload electrical sockets.
- Undertake regular maintenance of equipment.
- Be ‘Fire Aware’.
- Train staff to be ‘Fire Aware’.

Who can carry out a Fire Risk Assessment?

The Fire Precautions (Workplace) Regulations 1997 (amended) state that the person who carries out a Fire Risk Assessment shall be *competent* to do so.

Competency is not defined specifically in the Regulations, or in the Management of Health and Safety at Work Regulations 1999.

Competency means having a level of knowledge and experience which is relevant to the task in hand.

Many fire authorities, who enforce the Fire Precautions (Workplace) Regulations 1997 (amended) do not advocate that consultants are employed to carry out complicated assessments.

A Fire Risk Assessment is a logical, practical review of the likelihood of a fire starting in the premises and the consequences of such a fire. Someone who has good knowledge of the work activities and the layout of the building, together with some knowledge of what causes a fire to happen, would be best placed to carry out a Fire Risk Assessment.

What sort of fire hazards need to be considered?

Consider any significant fire hazards in the room or area under review:

- combustible materials, e.g. large quantities of paper, combustible fabrics or plastics
- flammable substances, e.g. paints, thinners, chemicals, flammable gases, aerosol cans
- ignition sources, e.g. naked flames, sparks, portable heaters, smoking materials, Hot Works equipment.

Do not forget to consider materials which might smoulder and produce quantities of smoke. Also, consider anything which might be able to give off toxic fumes.

Consider also the type of insulation involved or used in cavities, roof voids, etc. Combustible material may not always be visible, e.g. hidden cables in wall cavities.

What sort of structural features are important to consider when carrying out a Fire Risk Assessment?

Fire, smoke, heat and fumes can travel rapidly through a building if it is not restricted by fire protection and compartmentation.

Any part of a building which has open areas, open staircases, etc. will be more vulnerable to the risk of fire should one start.

Openings in walls, large voids above ceilings and below floors allow a fire to spread rapidly. Large voids also usually contribute extra ventilation, thereby adding more oxygen to the fire.

A method of fire prevention is to use fire-resistant materials and to design buildings so that fire will not travel from one area to another.

Any opportunity for a fire to spread through the building must be noted on the Fire Risk Assessment.

What are some of the factors to consider when assessing existing control measures for managing fire safety?

Many premises and employers already have some level of fire safety management in place and the Fire Precautions (Workplace) Regulations 1997 (amended) were not intended to add an especially heavy burden onto employers.

Existing control measures must be reviewed and the following are examples of what to look for:

- the likely spread of fire
- the likelihood of fire starting

- the number of occupiers of the area
- the use and activity undertaken
- the time available for escape
- the means of escape
- the clarity of the escape plan
- effectiveness of signage
- how the fire alarm is raised
- can the alarm be heard by everyone
- travel distances
- number and widths of exits
- condition of corridors
- storage and obstructions
- inner rooms and dead ends
- type and access to staircases
- openings, voids, etc. within the building
- type of fire doors
- use of panic bolts
- unobstructed fire doors
- intumescent strips
- well fitting fire doors
- propped open fire doors
- type of fire alarm
- location, number and condition of fire extinguishers
- display of fire safety notices
- emergency lighting
- maintenance and testing of fire alarm break glass points
- installation of sprinklers
- location and condition of smoke detectors
- use of heat detectors
- adequate lighting in an evacuation
- training of employees
- practised fire drills
- general housekeeping
- management of contractors
- use of Hot Works Permits
- control of smoking

- fire safety checks
- provision for managing the safety of people with disabilities
- special conditions, e.g. storage of flammable substances
- storage of combustible materials near a heat source.

The best Fire Risk Assessments are 'site specific' — review and inspect your *own* workplace.

Checklist

What should a Fire Risk Assessment cover?

- Identification of hazards.
- Sources of ignition.
- Identification of persons at risk from fire.
- Means of escape from the building.
- Fire warning systems.
- Fire-fighting facilities.
- Identification of fire safety procedures, i.e. emergency procedures.
- Review of the controls in place and recommendations for improvements where necessary.

The following pages contain a number of different types of Fire Risk Assessment formats.

FIRE RISK ASSESSMENT

Name of premises: ______________________________

What particular area are you reviewing for this Fire Risk Assessment?

What activity, practice, operation, etc. are you particularly reviewing for this Fire Risk Assessment?

What ignition sources have you identified?

What sources of fuel have you identified?

Are there any 'extra' sources of oxygen, or will mechanical ventilation increase oxygen levels?

Does anyone do anything that will increase the risk of a fire starting?

If a fire were to start, who would be at risk?

Would anyone be at any extra or special risk, or would any injuries/consequences of the fire be increased?

What precautions are currently in place to reduce the likelihood of a fire occurring, or to deal with it/control it if a fire did start (e.g. checks, alarms, emergency procedures, etc.)?

What other precautions need to be taken, if any? Does anything need to be done to improve existing fire precautions?

How will the information in this Risk Assessment be communicated to staff? Will any staff training take place?

Is there anything else that you think needs to be recorded on this Risk Assessment?

After having identified the hazards and risks of a fire starting and after considering all the procedures you *currently* have in place, do you consider the risk to *life safety* of either staff or customers (including any contractors, visitors, etc.) to be:

High ☐ Medium/normal ☐ Low ☐

If risks to life safety are very likely or possible, steps MUST be taken to improve fire safety.

If you implement the other, additional measures identified in this Fire Risk Assessment, will risk to *life safety* of either staff or customers (plus others) be:

High ☐ Medium/normal ☐ Low ☐

If risks to life safety are possible or very likely, then greater control measures MUST be implemented.

Risk Assessment completed by: ______________________________

Date: ______________________________

Fire Risk Assessment needs a review on: ______________________________

FIRE RISK ASSESSMENT

Name and address of premises: ______________________________

____

____

Owner/Employer/Person in Control: ______________________________

Contact details: ______________________________

Date of Risk Assessment: ______________________________

Completed by: ______________________________

Use of premises/area under review: ______________________________

Identification of fire hazards	High	Medium	Low

Identification and location of those at risk
Evaluation of the risks
Significant findings
Actions taken to reduce/remove risks
Residual Risk Assessment High ☐ Medium/normal ☐ Low ☐
Review of Risk Assessment: Under what circumstances: How often:

FIRE RISK ASSESSMENT

Name and address of premises:				
Area/room/floor under assessment:				
Name of person completing assessment:			Date of assessment:	
Fire hazard	**People at risk**	**Existing control measures**	**Proposed action to be taken**	**Date action taken and by whom (include signature)**

FIRE SAFETY RISK ASSESSMENT

Name of premises:			
Date of Assessment:			
Name of person carrying out Assessment:			
Area of premises being Assessed:			
Identification of Fire Safety Hazards	Yes	No	Don't know
Combustible materials, furnishings, e.g. carpets, curtains, seat coverings			
Smokers and discarded smoking materials			
Electric appliances, e.g. portable equipment			
Storage of chemicals, flammable aerosols, etc.			
Portable heaters, LPG, flambé lamps			
Cooking equipment, ducts, gas pipes			
Accumulations of waste, including cooking oil			
Obstruction of vents or cooling systems			
Building works			
Items stored too near to heat sources			
Obstruction of fire exit routes or fire doors propped open			

Comments (Please describe anything highlighted 'yes' above)			
Furnishings and fittings throughout the premises, including public areas, bars, restaurants, event rooms, bedrooms and offices. Smokers are permitted within the premises and smoking waste is controlled by discarding smoking materials into designated ashbins. Electrical appliances including televisions, kettles, irons, hairdryers, glass washers, microwaves, other kitchen appliances and computers. Chemical storage for cleaning products and storage of CO_2. Flambé lamps used within the restaurants. Cooking equipment includes gas oven stoves, grills and deep fat fryers in all kitchens and canopy extract systems. Accumulation of waste in bar, kitchen, housekeeping and refuse area including discarded smoking material, paper waste and collection of waste cooking oil.			
Who might be at risk from the above hazards: • customers • staff • disabled people • contractors?			
Are any people *particularly* at risk if a fire should break out, e.g. people working alone in plant rooms, visitors, etc.? Describe how and why they will be at risk:			

WHAT FIRE SAFETY CONTROL MEASURES ARE ALREADY IN PLACE

CONTROL MEASURE	DESCRIBE
Fire alarm — warning	
Fire alarm — detection	
Escape lighting	
Escape routes	
Emergency signage	
Fire extinguishers	
Fire blanket	
Testing of the system — what records are kept?	
Fire escape routes — how are they maintained clear of obstructions, how often are they checked, etc.?	

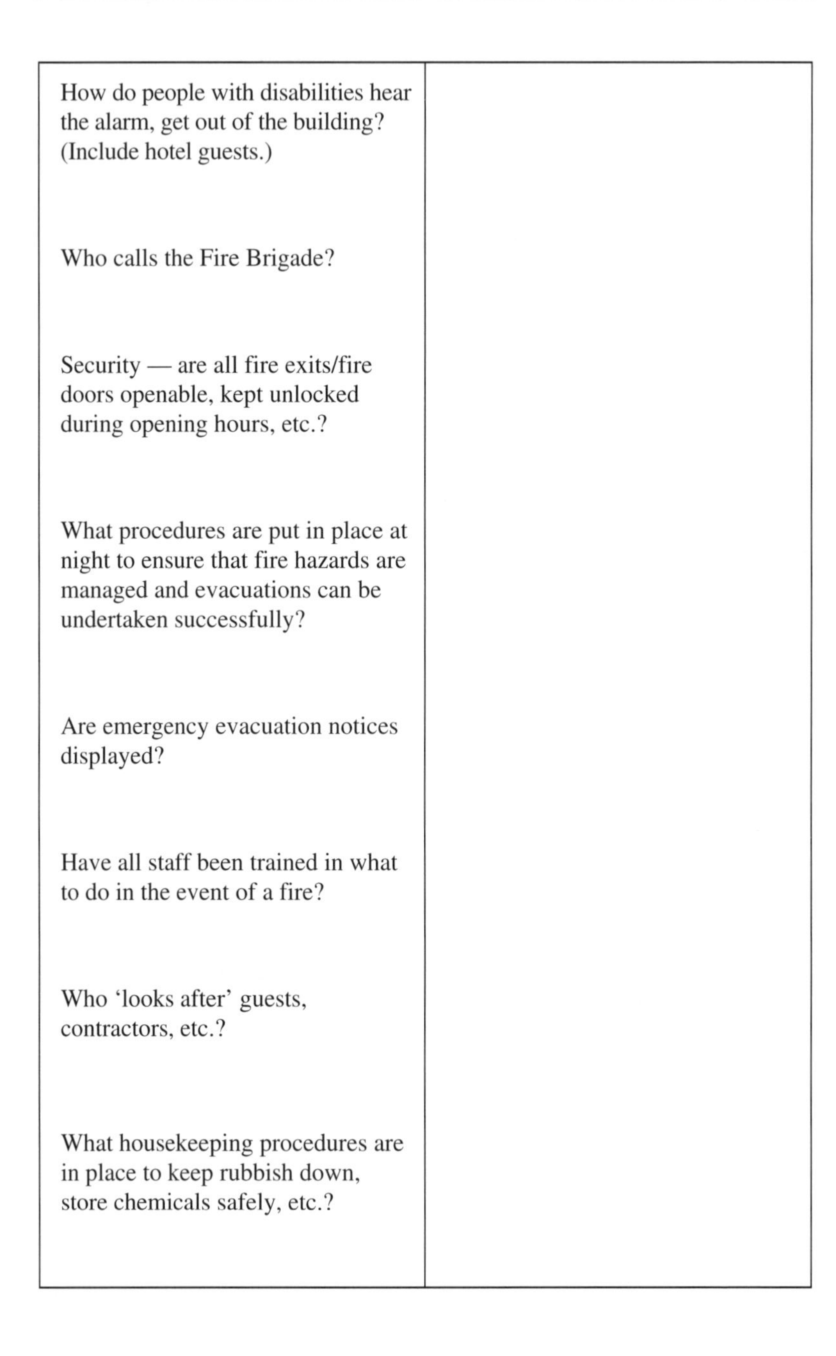

How do people with disabilities hear the alarm, get out of the building? (Include hotel guests.)	
Who calls the Fire Brigade?	
Security — are all fire exits/fire doors openable, kept unlocked during opening hours, etc.?	
What procedures are put in place at night to ensure that fire hazards are managed and evacuations can be undertaken successfully?	
Are emergency evacuation notices displayed?	
Have all staff been trained in what to do in the event of a fire?	
Who 'looks after' guests, contractors, etc.?	
What housekeeping procedures are in place to keep rubbish down, store chemicals safely, etc.?	

Is equipment regularly maintained to ensure that it works effectively, thus reducing the risk of electrical failure?	
FURTHER COMMENTS	
WHAT ADDITIONAL CONTROL MEASURES ARE NEEDED?	
Risk Assessment review date: Risk Assessment to be reviewed annually or whenever circumstances change that necessitate reassessment.	

EXISITING RISK RATING (WITH CURRENT CONTROLS)		
Low____	Medium____	High____
PROPOSED RISK RATING (WITH RECOMMENDED CONTROLS WITHIN THIS RISK ASSESSMENT)		
Low____	Medium____	High____

FIRE RISK ASSESSMENT

PART ONE

Location (describe area under review)
Department occupying area or description of activity
Occupiers/persons at risk Number of employees Number of customers (approx.) Number of others Comment on any extreme conditions for occupancy, eg. functions, banquets, sports competitions, etc.
Fire hazards Ignition sources (description of any likely sources of ignition, e.g. naked flames, heaters, smoking equipment, Hot Works, electrical faults, etc.)

<table>
<tr><td>Sources/types of combustible materials (describe levels of stock, packaging, chemicals, fuels, dust, etc. which could ignite)</td></tr>
<tr><td>Activities undertaken which could create hazards (describe any unusual or infrequent activity which occurs in the area which could contribute to combustion, etc. (e.g. Hot Works, contractors, etc.))</td></tr>
<tr><td>Structural condition of building (describe any known structural condition, defects, etc., which could create a hazard or cause spread of fire which could cause a risk to life safety of persons, not structural damage)</td></tr>
<tr><td>Persons at risk from fire (describe all persons whose safety could be at risk if a fire occurred in the location)</td></tr>
<tr><td>Risk Assessment without controls in place

High _____

Medium _____

Low _____</td></tr>
<tr><td>Additional comments on Part 1 — Fire Risk Assessment</td></tr>
</table>

PART TWO — CONTROL MEASURES

Means of escape (describe the means of escape from the location, including any further means of escape out of the building)
Structural precautions (describe any structural precautions taken to prevent the spread of fire or smoke through the building, e.g. 30 minutes/1 hour fire protection to structural beams, compartmentation, fire dampers, flame retardant materials, intumescent paint, intumescent strips/seals, smoke seals, etc.)
Fire detection measures (describe sprinklers, smoke and heat detectors, visual inspections, etc.)
Emergency lighting (describe level and location of emergency lighting)
Fire alarm system (describe methods of raising the alarm)
Means of fire fighting (describe fire extinguishers, sprinklers, oxygen reduction methods, etc. used to extinguish a fire when appropriate)

Hazard reduction methods (describe housekeeping, methods, reduction of combustible material, control of contractors, Hot Work Permits, reduction/elimination of ignition sources, etc.)
Fire precautions and training (describe management controls in place, e.g. fire wardens, weekly/daily fire safety checks and level and type of training for employees, etc.)
Maintenance, testing and records (describe what is done to maintain fire precautions, what tests are carried out, e.g. fire alarms, sprinklers, etc. and what records are kept and where)
Other control measures applicable to the area

PART THREE

What additional control measures are needed to protect the safety of all persons, i.e. employees, guests, customers, visitors and contractors using the building. What, if any, of the control measures listed in Part Two need to be improved and to what degree and by when?

Control measure	By whom	When

PART FOUR

Confirmation that the above remedial measures have been completed.

PART FIVE

Residual fire risk:

High____ Medium/normal____ Low_____

EXAMPLE OF COMPLETED FIRE RISK ASSESSMENT

Name of premises: Anywhere Hotel			
Date of Assessment: June 2002			
Name of person carrying out Assessment: Perry Scott Nash Associates Limited			
Area of premises being Assessed: Restaurant, Kitchen, Bar, Champagne Bar, Production Kitchen, Pastry Kitchen and Bedrooms.			
Identification of Fire Safety Hazards	Yes	No	Don't know
Combustible materials, furnishings, e.g. carpets, curtains, seat coverings	✓		
Smokers and discarded smoking materials	✓		
Electric appliances, e.g. portable equipment	✓		
Storage of chemicals, flammable aerosols, etc.	✓		
Portable heaters, LPG, flambé lamps	✓		
Cooking equipment, ducts, gas pipes	✓		
Accumulations of waste, including cooking oil	✓		
Obstruction of vents or cooling systems		✗	
Building works		✗	
Items stored too near to heat sources		✗	
Obstruction of fire exit routes or fire doors propped open		✗	

Comments (Please describe anything highlighted 'yes' above)			
Furnishings and fittings throughout the premises, including public areas, bars, restaurants, event rooms, bedrooms and offices. Smokers are permitted within the premises and smoking waste is controlled by discarding smoking materials into designated ashbins. Electrical appliances including televisions, kettles, irons, hairdryers, glass washers, microwaves, other kitchen appliances and computers. Chemical storage for cleaning products and storage of CO_2. Flambé lamps used within the restaurants. Cooking equipment includes gas oven stoves, grills and deep fat fryers in all kitchens and canopy extract systems. Accumulation of waste in bar, kitchen, housekeeping and refuse area including discarded smoking material, paper waste and collection of waste cooking oil.			
Who might be at risk from the above hazards: customers: • to pub • to hotel staff disabled people contractors?	 ✓ ✓ ✓ ✓ ✓ ✓		

Are any people *particularly* at risk if a fire should break out, e.g. people working alone in plant rooms, visitors, etc.? Describe how and why they will be at risk: Controls are in place to ensure contractors and visitors are accounted for. Procedure in place indicates person's name, time of arrival, area of work, type of work, expected time of work/visit, and each person signs out before leaving. However, some contractors might leave the hotel without informing security. Disabled/special needs persons accommodated on the first and second floor. In the event of a fire the lift cannot be used and only means of escape is via staircase. Special needs rooms — 161, 167, 205 and 207; all front desk staff are aware of occupants' needs.			

WHAT FIRE SAFETY CONTROL MEASURES ARE ALREADY IN PLACE

CONTROL MEASURE	DESCRIBE
Fire alarm — warning	Alarm bell activated either by fire call alarm points, automatic smoke and heat detection, Stratos — high sensitivity photo electric smoke detector OR by activating the main fire alarm panel. A continuous ringing throughout the premises, which can only be silenced at the alarm panel. All fire call alarm points are clearly signed throughout the premises.
Fire alarm — detection	Automatic heat and smoke detectors located throughout the premises. The Hotel is a grade two listed building and the Stratos — high sensitivity smoke detectors are situated within the fabric of the building.
Escape lighting	Emergency lighting luminaire units installed over all fire exit routes indicating escape routes and also installed within escape routes located from the ceiling using directional signage to define escape route.
Escape routes	From lower and upper basement via internal metal staircase to ground level. From ground, first, second, third, forth, fifth and sixth floor areas via internal routes including metal staircase or by external metal staircase. All routes are segregated by automatic fire doors, which close in the event of a fire to prevent fire spread.

Emergency signage	Directional signage clearly defines escape routes. Staff fire action route signage located throughout the premises with pictorial and written instruction.
Fire extinguishers	Fire extinguishers and fire hoses located on every level of the hotel. Fire extinguishers include water, foam, powder and CO_2 (refer to Appendix 1 for type of fire extinguisher and location) also (refer to Appendix 2 for fire hose location).
Fire blanket	Nine fire blankets located within all kitchens at the Hotel. (Refer to Appendix 3 for fire blanket location.)
Testing of the system — what records are kept?	Fire call alarm points are tested every Friday at 16:00. Monthly testing of fire extinguishers, fire exit/escape and emergency lighting. A competent contractor undertakes annual tests of fire extinguishers, last tested in January 2002, Ansul system last tested 30 April 2002, lighting is ongoing annually (refer to Appendix 4 for electrical testing). All monthly and annual tests are recorded and appropriate certification is kept.
Fire escape routes — how are they maintained clear of obstructions, how often are they checked, etc.?	Staff members are informed to keep all fire escape routes clear including corridors and fire exit doors. Daily and nightly checks are carried out to ensure routes are not obstructed. CCTV allows security to permanently check routes. Any obstructions are logged and removed.

How do people with disabilities hear the alarm, get out of the building? (Include hotel guests.)	Front desk staff keep details of guests with special needs. In the event of a fire, a disabled person would be assisted down via the staircase if safe to do so. Staff would assist and guide other guests/patrons to the assembly point, on route to leaving the premises.
Who calls the Fire Brigade?	Management and senior staff are responsible for formulating the emergency plan in the event of a fire. Front desk staff would call the Fire Brigade under instruction.
Security — are all fire exits/fire doors openable, kept unlocked during opening hours, etc.?	All fire exits/fire doors have push bar operation or turn knobs to fully open doors in the event of a fire. No fire door is locked or requires a key to open it in the event of a fire.
What procedures are put in place at night to ensure that fire hazards are managed and evacuations can be undertaken successfully?	Night staff undertake checks throughout the premises to include that fire doors are closed, corridors/fire routes are not obstructed and for signs of fire. Currently, no formal procedure implemented, however any concerns are noted on the night check sheet and, if necessary, a digital photo is taken for evidence. This information is then given to security during handover. All night staff receive three-monthly fire safety training to include evacuation.

Are emergency evacuation notices displayed?	Each room has an evacuation notice with plan and instructions for the event of a fire; this is displayed within a frame on the back of the bedroom door. Staff fire action route signage is located throughout the premises with pictorial and written instructions to include: i) operating alarm call point ii) calling the Fire Brigade iii) fire fighting iv) evacuation v) assembly point.
Have all staff been trained in what to do in the event of a fire?	All new staff receive induction fire safety training which is carried out every Monday. All daytime staff receive refresher fire safety training every six months and all night staff receive refresher fire safety training every three months. Evacuation training is carried out every six months with a full evacuation of the hotel. All staff sign for training and records are kept by the HR Department.
Who 'looks after' guests, contractors, etc.?	Management and staff look after guests in the event of a fire. A full occupation list is printed off and front desk staff check guests at the assembly point. Security look after contractors and visitors, full formal procedure of checking in and out of contractors and visitors.
What housekeeping procedures are in place to keep rubbish down, store chemicals safely, etc.?	All chemicals are stored in locked cupboards with full COSHH information for each chemical. All waste is removed to designated refuse area with frequent collection to remove build-up of waste.

Is equipment regularly maintained to ensure that it works effectively, thus reducing the risk of electrical failure?	Yes. Portable appliance testing (PAT) carried out annually. Maintenance record all tests carried out. All fixed equipment is tested by a competent person every two to three years and recorded/certificated with the Maintenance Department. Staff remain vigilant and report any defects to heads of department for action.
FURTHER COMMENTS	

WHAT ADDITIONAL CONTROL MEASURES ARE NEEDED?

1. Provide signage to emergency stop buttons situated within the kitchen to indicate gas, electricity or both. (**7 days**)
2. Remove cardboard boxes and other items stored within the electricity cupboard situated in the Bar and ensure that regular checks are made to prevent overstocked items obstructing electrical panels. (**Immediate**)
3. Provide designated ashbin within the staff canteen for discarding of waste smoking materials. (**Immediate**)
4. Provide correct signage and fix fire extinguishers A49 and A50 within the electrical plant room situated on lower basement. (**Immediate**)
5. Replace directional fire exit sign situated within the laundry corridor. It is recommended that a plastic sign be fixed to the wall higher than the top of contract laundry trolleys to prevent damage to the sign. (**7 days**)
6. Replace fire blanket number one in the staff canteen kitchen as this was found to be greasy and dirty during the Fire Risk Assessment audit. Ensure that all fire blankets are checked for condition and replace when necessary. (**7 days**)
7. It is unclear if the fire hoses have recently been checked by a competent person. Check with Fire Protection who carried out the recent examination during January 2002 and obtain records. (**7 days**)

8. Ensure all fire doors leading onto escape routes are fitted with intumescent strips along the edges of the door. The majority of fire doors have been fitted with intumescent strips but it was noted that several kitchen fire doors had not — investigate and complete works. (**7 days**)
9. It is strongly recommended that a fire-proof safe be installed to store all fire safety records, to prevent damage or loss in the event of a fire. (**1 month**)
10. Provide emergency button within the steam room situated in the Train gym. It is understood that the works are in hand — awaiting installation of emergency button.
11. Implement formal documentation for night checks carried out indicating time, location and person carrying out checks. This may be paper- or computer-based documentation. It is recommended that the night fire safety checks are carried out half hourly between midnight and 07:00, to ensure that all areas of the premises are checked. (**7 days**)
12. The metal fire staircase situated at the fire exit, has been in-filled with wooden blocks to prevent a trip hazard. The in-fills should be metal to provide integrity to the fire escape route. Investigate and replace the in-fills with metal to match existing fabrication. (**1 month**)
13. Review Fire Risk Assessment yearly or when any changes are made to fire-fighting equipment/detection or internal procedures to ensure that it reflects current operational procedures within the Hotel. (**Ongoing**)

Risk Assessment review date:

Risk Assessment to be reviewed annually or whenever circumstances change that necessitate reassessment.

EXISITING RISK RATING (WITH CURRENT CONTROLS)

Low ✓ Medium____ High____

PROPOSED RISK RATING (WITH RECOMMENDED CONTROLS WITHIN THIS RISK ASSESSMENT)		
Low ✓	Medium____	High____

References

Fire Precautions (Workplace) Regulations 1997 (amended).

Management of Health and Safety at Work Regulations 1999.

Five steps to Risk Assessment: HSE: INDG 163.

Fire safety — an employer's guide: HSE Books, Home Office.

4

Fire Certificates

What is a Fire Certificate?

A Fire Certificate is a document issued by the Fire Authority in respect of a certain category of premises in which fire safety has to be specifically managed because of the potential risk of serious personal injury to occupiers, visitors, etc. should a fire break out.

The Fire Certificate gives requirements for fire precautions, fire protections, fire monitoring and management and employee training.

What types of premises need a Fire Certificate?

Fire Certificates are required for the following premises which have been 'designated' by Parliamentary Orders:

- factories, offices, shops and railway premises employing *more than* 20 people (or more than 10 employed elsewhere than on the ground floor)
- premises where certain flammable substances are kept
- premises used as a hotel or boarding house where sleeping accommodation is provided for more than *six* persons, being

staff or guests, or sleeping accommodation is provided above the first floor or below ground floor for guests or staff.

Certain premises will also need a Fire Certificate if they are designated as 'special premises'.

The Fire Certificates (Special Premises) Regulations 1976 deal with very high risk premises, e.g. oil refineries, and these certificates are issued by the Health and Safety Executive.

Who should be counted as an employee?

A Fire Certificate applies to a building in which people are employed to work.

In order to calculate the number of employees in the building, those employed in all areas of the building must be counted. This means that if a building has several occupiers or employers of small numbers of staff, it is the total number of people which has to be calculated.

A person at work is:

- an individual who works under a contract of employment or apprenticeship
- an individual who works for gain or reward otherwise than under a contract of employment or apprenticeship, whether or not he employs other persons
- a person receiving training under any of the Employment Acts.

Employers must take into the calculation any contract, agency or self-employed persons.

How should part-time employees be counted?

The Fire Precautions (Factories, Shops and Railway Premises) Order 1989 does not differentiate between full-time and part-time employees.

The actual wording for calculating the number of employees in respect of a Fire Certificate is 'persons at work at any one time'.

So, what counts is the number of people, i.e. employees, at work in the building at any one time. The payroll may have over 20 people listed but if they are not all at work at the same time the premises will not need a Fire Certificate, or if not more than 10 of them are at work on the floors other than ground.

Two part-timers at work at the same time do *not* count as one employee for the purposes of the Fire Certificate.

Is a Fire Certificate necessary if staff accommodation is provided?

In many instances, the answer will be yes.

The Fire Precautions (Hotels and Boarding Houses) Order 1972 sets out the criteria for which residential or sleeping accommodation needs Fire Certificates.

If premises are used for providing sleeping accommodation for *more than* six staff members, then a Fire Certificate will be required.

If sleeping accommodation is provided for fewer than six staff, but the accommodation is provided on any floor above the first floor of the building which constitutes or comprises the premises then a Fire Certificate will be needed.

Sleeping accommodation provided below ground level, i.e. at basement level, will also require a Fire Certificate irrespective of the number of staff being catered for.

Who should apply for a Fire Certificate?

In the case of offices, shops and railway premises which meet the requirements of needing a Fire Certificate, application should normally be made by the occupier of the premises.

This may be difficult in multi-occupied buildings and the duty to apply can fall to the owner.

The owners of premises have a duty to apply for the Fire Certificate for the following types of premises:

- premises held under a lease agreement or license to occupy where all parts of the building are owned by one person
- premises in multi-ownership and multi-occupancy — all owners should apply — often this will be through the Managing Agents.

How is a Fire Certificate applied for?

Application for a Fire Certificate is normally to the Local Fire Authority.

A specific Fire Certificate application form is available, known as FP1.

Applications for Fire Certificates are not necessarily dealt with immediately due to the large numbers which Authorities receive and the resources available in their Fire Protection divisions.

Most Fire Authorities will acknowledge the application and will process it in due course.

As long as the Fire Certificate has been applied for, the owner or occupier of the building will not be breaking the law.

However, an application 'in the pipeline' does not negate responsibilities for managing fire safety in the workplace.

The Fire Precautions (Workplace) Regulations 1997 (amended) set down employer responsibilities for managing fire safety irrespective of whether the premises has a Fire Certificate or not.

What are 'Interim Duties' under the Fire Precautions Act 1971?

The Fire Precautions Act 1971 requires owners or occupiers of premises which require a Fire Certificate to carry out 'interim duties' in the period between applying for a Fire Certificate and one being formally issued by the Fire Authority.

The Fire Authority will eventually carry out an inspection of the premises in respect of the Fire Certificate application but, until that occurs, they expect owners or occupiers to:

- ensure that any existing means of escape in case of fire provided within the premises can be safely and effectively used when people are on the premises
- ensure that any existing fire-fighting equipment provided within the premises is maintained in efficient working order
- ensure that all employees in the premises receive instruction or training in what to do in case of fire.

Can premises be exempted from requiring a Fire Certificate?

Yes, at the discretion of the Fire Authority.

The Fire Precautions Act 1971 allows for Fire Authorities to exempt certain buildings from being issued with a Fire Certificate if the Fire Authority believe them to be 'low risk'.

An application for a Fire Certificate must be made in the usual way.

The Fire Authority will either issue an Exemption Notice once they have assessed the application and plans, or once they have inspected the premises. Exemption Notices are purely at the discretion of the Authority.

If an Exemption Notice is issued it will state:

- that an exemption from the requirements to have a Fire Certificate has been granted for a particular use or uses of the premises, and
- may specify the greatest number of persons who can safely be in the premises at any one time.

What does a Fire Certificate cover?

The Fire Certificate will contain the following information about the premises to which it relates:

- the particular use or uses of the premises which it covers
- the means of escape in case of fire provided for the premises
- the means for ensuring that the means of escape provided can be safely and effectively used at all relevant times
- type and numbers of fire safety signs, emergency lighting, fire-resisting doors, fire protection measures, etc.
- types and numbers of fire-fighting equipment required, e.g. fire extinguishers, sprinklers, hose reels, etc.
- the type and means of giving warning in respect of a fire, e.g. fire alarm systems.

The above are mandatory inclusions in a Fire Certificate. In addition, the Fire Authority may include any or all of the following:

- maintenance of the means of escape provided
- methods for keeping means of escape free from obstruction
- maintenance of fire-fighting equipment
- maintenance of other fire precautions
- training of employees in what to do in the event of a fire and emergencies
- requirement for fire drills, etc.
- requirement for training records
- limitations as to maximum number of persons who can be on the premises or in certain parts of it
- any other relevant fire precautions, e.g. fire retardant materials or smoke control measures.

Fire Certificates usually include a plan of the building. The plan is annotated with means of escape, fire alarm call points, positions for fire-fighting equipment, type and location of fire safety signs, fire doors and fire-resistant construction.

What is the legal status of a Fire Certificate?

The recipient of a Fire Certificate has a legal duty to comply with the requirements of the Fire Certificate.

Failure to comply with a Certificate is an offence.

Anyone issued with a Fire Certificate has a duty to:

- keep the Fire Certificate in the relevant building
- maintain the fire safety provisions, and any other requirements, *precisely* as itemised in the Certificate
- give notice to the Fire Authority of any *material* change to the building by way of extensions or structural alterations
- give notice to the Fire Authority of any material internal alterations, or alterations to furniture layout, equipment use, etc.
- inform the Fire Authority if any quantities of flammable materials will be stored on the premises.

How should 'material changes' to the building be notified to the Fire Authority?

Anything which could materially interfere with the fire precautions within the building must be notified to the Fire Authority. New building works, structural alterations, etc. could impede means of escape and compromise the safety of occupiers. Means of escape, for instance, is often determined by 'distance of travel' and whether one or more exit routes are available.

If the Fire Authority deem that the changes and alterations proposed to the premises materially alter the provisions for fire safety, they will issue a 'steps to be taken' letter or notice. This will itemise what steps need to be taken to ensure that fire precautions in the building are maintained.

Failure to comply with the additional requirements may result in the Fire Authority cancelling the Fire Certificates. Failure to have a current Fire Certificate, or to have applied for one, is an offence.

If the recommendations of the Fire Authority are followed, the Authority will issue an Amended Fire Certificate.

What are special premises under the Fire Precautions Act 1971?

Special premises are premises which operate hazardous processes or which store, manufacture or otherwise handle certain specified hazardous materials in quantities which present a significant risk to health and safety for employees and members of the public.

The Health and Safety Executive issue the Fire Certificate.

The number of persons employed in a 'special premises' is irrelevant and a Fire Certificate may be required for only a few members of staff.

The HSE have powers to grant exemptions to certain premises if they deem it appropriate to do so.

The 'responsible person' is deemed to be the person who must apply for a Fire Certificate.

Once a Fire Certificate for special premises has been granted, a copy must be posted in a significant location so that it can easily be read by persons who might be affected by any of the provisions of the Certificate.

Generally, special premises are those which:

- keep highly flammable materials or substances under pressure
- store LPG in quantities in excess of 100 tonnes, unless the gas is kept as fuel
- store liquefied natural gas in quantities greater than 100 tonnes except when kept as fuel
- store methyl acetylene gas in quantities greater than 100 tonnes
- manufacture oxygen and store liquid oxygen in excess of 135 tonnes
- store chlorine in excess of 50 tonnes unless stored for water purification

- store a range of highly flammable substances in various quantities such as:
  - ethylene oxide
  - acrylonitrile
  - ethylene
  - phosgene
  - carbon disulphide
  - hydrogen cyanide
  - propylene
  - other flammable liquids
- store explosives
- arc associated with mines
- are associated with nuclear installations
- are associated with radioactivity and ionising radiation
- are used as temporary buildings on construction or engineering sites.

Reference

Fire Precautions Act 1971.

5

Fire precautions and defence systems

What is meant by the term 'fire precautions'?

Fire precautions usually refer to the protection measures and other procedures which are taken to protect people from the consequences of a fire.

Employers have duties to protect their employees and others from the hazards and risks generated by their business, including fire. The measures which they adopt are collectively called 'fire precautions'.

The Government's *Fire safety — an employer's guide* covers the following subjects in Part 3, which is entitled 'Further Guidance on Fire Precautions':

- maintenance of plant and equipment
- storage and use of flammable equipment
- flammable liquids
- work processes involving heat
- Hot Works
- electrical equipment
- heating appliances
- smoking
- building and maintenance work
- flammable rubbish or waste
- arson

- fire detection
- means of escape
- fire warning
- fire detection
- fire evacuation
- emergency lighting
- smoke and heat detection
- fire-fighting equipment
- fire signage.

While all fire precautions are important, are any more important than others?

Yes. The main and most important fire precautions in any building, premises or outdoor arena is the means of escape in case of fire.

Many people die in fires because they have been unable to escape the fire, or because the smoke and fumes have 'contaminated' the means of escape, preventing them from using it.

The Fire Authority are especially interested in means of escape when they carry out their statutory inspections.

The other most important aspect of fire precautions is actually 'fire prevention' — why do fires start in the first place? Remember the first 'law' of risk assessment — eliminate the hazard!

What is meant by the term 'means of escape'?

The term 'means of escape' in fire safety refers to the means which are provided in a building for occupants to leave the building in an emergency, quickly and safely without unnecessary assistance.

Means of escape are usually alternative exits from a building which have been designed to be accessed quickly from an occupied area.

Means of escape should lead to a place of safety, should be easy to identify and safe to use.

Means of escape should allow escape from a building in more than one direction and in a direction opposite to that of the main entrance.

Are there any exit routes which are not suitable means of escape?

Any escape route which could be hazardous to use would be unsuitable. In particular, the following are considered to be *unsuitable* as means of escape:

- lifts
- portable ladders
- long spiral staircases
- escalators
- 'throw out' devices or escape ropes, etc.

Why is it not acceptable to use lifts as a means of escape?

Lifts are unsuitable as a means of escape because:

- the power may be disconnected because of the fire and occupants of the lift car could be trapped in the lift shaft in a burning building
- lifts could stop and disgorge occupants onto the floor on which the fire is raging. Lift occupants will not know which floor has the fire and could walk into it — becoming trapped.

There are some designs of lifts which can be used in a fire and these are known as 'fire fighter' lifts. They are designed to run on completely separate power systems and are installed for the use of fire fighters so that they can get to the scene of a fire in a tall building quickly.

What is a 'place of safety'?

A place of safety is a place in which a person is no longer at risk from a fire.

A place of safety must be in the open air and must be a place beyond the building.

Means of escape take people to a place of safety, i.e. to a location away from the building where they will not be at risk from heat, fumes, smoke, collapsing structures, etc.

It is not adequate for occupants to discharge from the building into an adjoining alleyway, small yard, inner courtyard, rear garden, etc. from which there is no escape and where occupants would still be at risk.

Is a place of safety the same as an assembly point?

Not necessarily so. Means of escape lead people *to* a place of safety, i.e. an area in the open air beyond the building.

An assembly point is a location at which people are requested to congregate so that a roll call can be made to account for all occupants, employees, etc.

What are the key principles involved in designing means of escape?

The first fundamental principle of a means of escape is that people should be able to turn their back to a fire and leave the building to a place of safety.

Buildings should preferably have two means of exit but smaller buildings may not.

Dead ends should be avoided as people may get to the end of the corridor and have no other way to go. This could mean that they would have to turn back towards the fire to exit the building.

Means of escape should be wide enough to take the occupants of the area without crushing and overcrowding.

Exits from areas of the building should be separate and independent and not so close together that they really do not present an alternative.

Exit routes from upper floors will obviously not lead initially to an outside place of safety. In this instance, an exit route must lead to a place of 'relative safety'.

What is a place of 'relative safety'?

Where there is a considerable distance between an exit route and the final, external place of safety, it is necessary to provide an intermediate place of safety — known as the place of relative safety. This will be a location where people could rest for a few seconds to gather their bearings, catch their breath or prepare themselves for the final descent to the final exit.

Multi-storey buildings have long distances from the upper floors to the ground exit routes and places of relative safety are provided on protected staircases, i.e. it is relatively safe to wait in the staircase pending the next stage of the exit.

What is meant by the term 'travel distance'?

Travel distance can be defined as the distance to be travelled from any part of the building to the nearest:

- final exit route and place of safety in the open air, away from the building
- door to a protected staircase or other place of relative safety
- door to an escape route located externally.

What are the legal requirements regarding distance of travel in buildings?

Distances of travel are laid down in the British Standards applicable to fire safety, namely BS 5588 Part II (formally Part 2) and also in the formal Code of Practices issued by the government on fire safety.

The distance of travel in a building depends on the design and use of the building and the assessment of the fire risk.

Single storey buildings

If two exit routes are provided this would be acceptable provided that the distance to either exit was not more than:

- 45 m in offices and factories
- 30 m in shops.

Exits must not be so sited that they are too close together, making them both inaccessible in the event of a fire.

In order to judge the effectiveness of exit routes it is recommended that the angle formed between the two doors be measured.

If the two doors form an angle of access between each other greater than 45° they will be deemed to be acceptable as alternative exits. If the angle formed is less than 45° they are too close together to give an effective alternative.

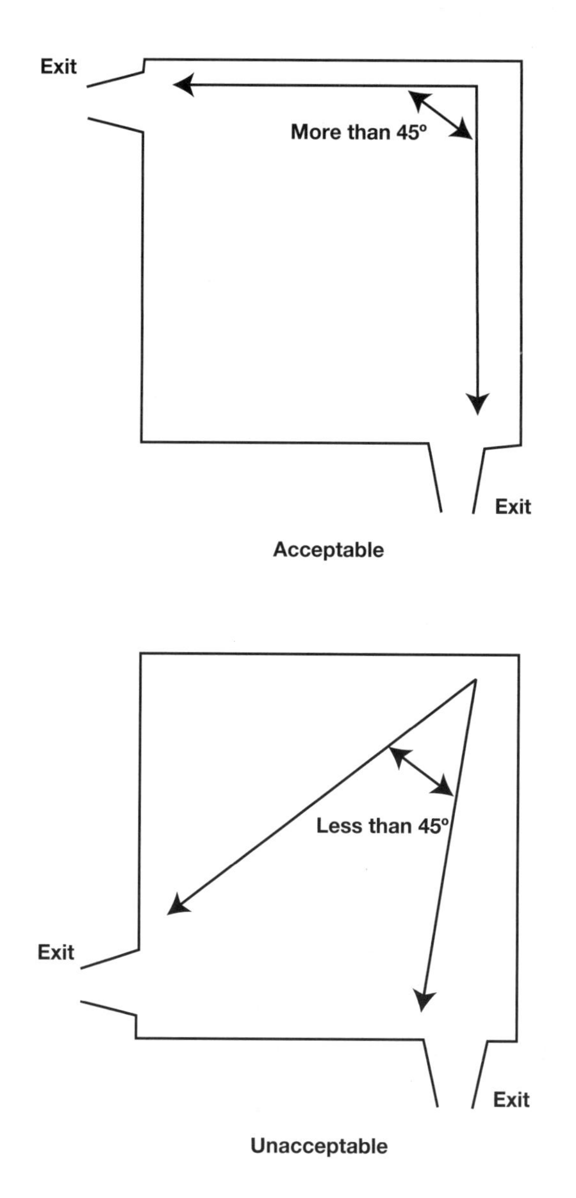
Exit
More than 45º
Exit
Acceptable
Less than 45º
Exit
Exit
Unacceptable

Where a floor is subdivided into offices or other rooms, corridors, etc. there is a maximum travel distance from the room to a 'room exit' of no more than 12 m.

In addition, the overall distance of travel to the final exit must not exceed 45 m in offices and factories and 30 m in shops.

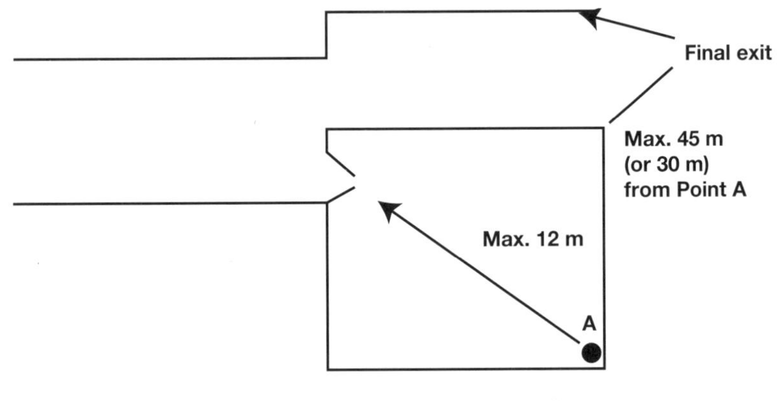

Single storey building

If only *one* exit is provided to a single storey building, the distance of travel to a final exit will be:

- a maximum of 18 m in offices and shops
- a maximum of 25 m in factories.

Multi-storied premises

Distance of travel criteria which relates to single storey buildings should be followed for multi-storied buildings.

Stairways and staircases serving upper and lower floors must really be 'protected', i.e. have some form of fire-resistant construction.

The distance of travel to a protected staircase can be:

- a maximum of 18 m in offices and shops
- a maximum of 25 m in factories.

The distance to be measured is from the furthest point of the floor to the entrance to the protected fire staircase.

If the staircase is *not* protected, then the distance of travel from the furthest point of the floor to the *final exit* route on the ground floor must be:

- a maximum of 18 m in offices and shops
- a maximum of 25 m in factories.

If a building is provided with *more than* one staircase, then the distance of travel to each staircase must be:

- in unprotected staircases:
  - 45 m in offices and factories
  - 30 m in shops

  provided that those distances lead to the *final exits*
- in protected staircases:
  - 45 m in offices and factories
  - 30 m in shops

  provided that those distances lead to the exit route leading into the protected staircase.

Do the distance of travel calculations vary from building to building?

The distance of travel calculations will be affected by the risk rating of the premises in respect of fire.

If fire risk and the danger to life safety is considered to be high then there will be a necessity to evacuate people from the building very quickly. Therefore, the shorter the distance of travel to a place of safety, or place of relative safety, the better the fire precautions for the building.

Travel distances may need to be reduced to:

- 12 m in high risk shops to a storey exit in one direction
- 25 m in high risk shops to a storey exit in more than one direction
- 6 m from within a room to the room exit in high risk shops or offices.

If premises have a Fire Certificate, the Fire Authority may, at their discretion, increase the distance of travel in low risk buildings to 60 m to the final exits or storey exits where there is more than one escape route.

How is the distance of travel measured?

Distance of travel should be measured as being the actual distance to be travelled between any point in the building and the nearest storey exit.

The distance should not be 'as the crow flies' but should measure the actual route to be taken and must include detours around fixtures and fittings, etc.

What is an access room?

Where the only means of escape from a room is through another room, the outer room is called an access room.

What is an inner room?

An inner room is a room which is entered via an outer room or access room.

An inner room must be under the control of the access room occupier.

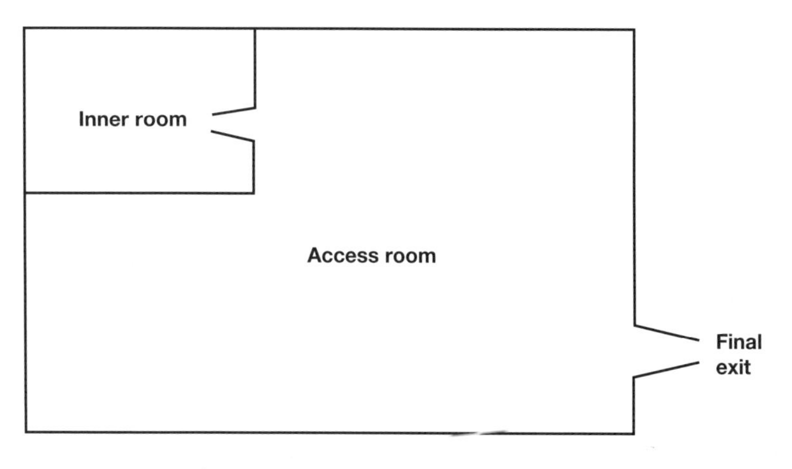

Inner room and access room

Why are these rooms important for fire safety?

Account must be taken of a fire starting in an access room and how someone in the inner room would be able to evacuate. A fire in the access room may restrict the ability of those in the inner room to leave the room.

Evacuation will be easier and quicker if the occupants of the inner room have clear visibility into the access room so that they can see either the beginnings of a fire or be made aware of the dangers by the people occupying the access room.

An inner room must therefore have suitable vision panels in any doors or partitions.

Occupants of the inner room must be able to detect the early stages of a fire.

An inner room should not be located off an access room which has a high fire risk rating, e.g. a store for hazardous or flammable chemicals, kitchen, etc.

The distance of travel from an inner room to the exit from the access room should not exceed 12 m.

Under certain circumstances, an inner room may not be provided with vision panels, e.g. sanitary accommodation or changing facilities.

Generally, these rooms would not be places of work and would not have occupancy of long duration.

It is *not* permissible to have an inner room entered via an access room which is itself entered via an access room.

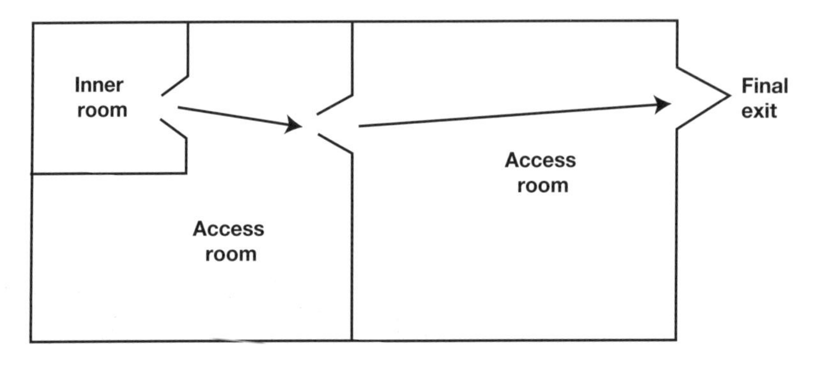

Unacceptable room layout

Do means of escape corridors have to be sub-divided so as to prevent the spread of fire?

Corridors exceeding 30 m in length in shops and 45 m in lengths in offices and factories should be sub-divided by fire doors to prevent the free travel of smoke, heat and toxic fumes.

The sub-division of corridors should be such that no individual length of corridor is common to more than one storey exit.

What is meant by the term 'the three stages of escape'?

Means of escape can be divided into three stages as follows:

Stage 1: Travel within a room
Stage 2: Horizontal travel to a storey exit or final exit
Stage 3: Vertical travel within a stairway to a final exit.

The three stages of escape is a useful tool when considering the complexities of means of escape. Like many complicated things, if they are broken down into manageable parts the task becomes easier to achieve.

So, when planning or checking means of escape think:

Stage 1: how do I get out of this room and where will I find myself?
Stage 2: how will I get out of this floor or main area and where will I find myself?
Stage 3: where are the staircases, where do they lead to and where will I find myself?

What is meant by the terms protected staircase and protected lobby?

A staircase is protected if it is enclosed by fire-resisting construction materials.

Fire-resisting materials are those which will withstand the onslaught of fire for a minimum of 30 minutes. Greater fire resistance is required in certain locations, e.g. public buildings.

A staircase is only protected when it is protected from fire in adjoining accommodation.

A protected staircase must discharge to a final exit or lead onto a protected route which then leads to the final exit.

A protected lobby is a fire-resisting enclosure providing access to a protected stairway via two sets of fire-resisting self-closing doors and into which no rooms open, other than toilets or lifts.

What is a protected route?

A fire protected route is one having a degree of protection from fire including walls, doors, partitions, ceilings and floors separating the route from the remainder of the building.

When might the principles of travel distances, protected routes, inner rooms, etc. need to be applied to existing buildings?

Generally, when an occupier, owner or employer is considering internal alterations and refurbishment works, they must take into consideration the fire safety requirements of the alterations.

Often, erecting internal partitions creates inner rooms leading off access rooms, or partitions will obstruct means of escape routes by increasing the travel distances.

It would be sensible to consider the existing means of escape before carrying out works and to plan the remedial works in relation to the principles of good fire safety.

The Fire Certificate must be consulted whenever alterations are proposed and any material alterations to the building must be reported to the Fire Authority as the Fire Certificate may need to be amended.

What are fire doors?

Doors which are fire-resisting and self-closing are usually referred to as 'fire doors'.

Fire doors are a vital part of a fire defence system as their purpose is to hold back fire and smoke.

They are normally provided to carry out one of two functions:

- to protect the integrity of a structural fire compartment by holding back fire and smoke
- to protect the means of escape for the occupiers of the building for a sufficient period of time for them to escape from the building.

Fire doors can give fire protection for periods ranging from 30 minutes to 4 hours.

Fire doors must be close fitting and constructed of material which withstands heat and flame for the period specified.

Fire doors are normally manufactured with intumescent strips around the door edge. This is, in effect, a smoke seal which expands in heat to fill the gaps between the door and door frame.

Fire doors must generally be kept shut otherwise their purpose is not fulfilled.

However, some fire doors can be designed to work with automatic door release systems which allow them to remain in the open position until an alarm or sensor triggers them to shut automatically.

Should all doors be fire doors?

It might be exceptionally good practice to have all doors designed to be fire doors, but it may be cost prohibitive.

Fire doors are needed on all fire escape routes, to rooms which have a high fire rating, e.g. plant rooms, electrical cupboards, store rooms, etc.

How do I know if a door is a fire door?

Generally, a fire door is thicker and heavier than a normal door and has an intumescent (brush or fuzzy) strip around its edge, or it is within a door frame which has an intumescent strip.

Often, the door stops are a minimum of 25 mm.

All fire doors should bear the mandatory sign:

Fire door — keep shut
Fire door — keep locked (for all store cupboards, electrical plant rooms, etc.).

It is sensible to be able to identify fire doors so that all employees know what and where they are and can assist in keeping them closed.

Is emergency lighting a form of fire protection?

Emergency lighting is not really a form of fire protection but it is a form of fire precaution.

All escape routes, including external ones, must have sufficient lighting for people to see their way out safely.

Emergency escape lighting may be required if lighting to escape routes is inadequate, i.e. the routes are illuminated with natural light or daylight, used at night or would be without light if the electrical power were interrupted.

Part of the Fire Risk Assessment will be to check the need for escape lighting to premises. Check all escape routes with the lights off. What can be seen? Are emergency exit routes visible? Is the exit route easy to negotiate and follow? Are there different levels, steps, etc. which people could fall down if they were unable to see them?

Sometimes borrowed lighting or street lighting may be sufficient to illuminate an escape route.

If the Fire Risk Assessment determines that the fire escape routes will be difficult to use without additional light them some form of emergency lighting will need to be provided.

Emergency lighting will need to be 'proportionate' to the risk and the number of people who will need to evacuate. In small premises with fewer than five employees and no members of the public, emergency lighting might be a powerful torch kept adjacent to the fire exit door or near the fire point.

In most other premises, emergency lighting should be provided which automatically comes on when the mains lighting circuit fails. This means that when the building falls into darkness because of a power failure, the emergency lighting circuit takes over and a level of light is generated which is sufficient to enable people to leave in safety.

Also, emergency lighting must be provided if this will help to shut down process machinery, etc. prior to an evacuation.

Emergency lighting should be sited to cover specific areas:

- intersections of corridors
- at each exit door

- near each staircase
- on each flight of stairs
- close to a change in floor level
- outside each final exit
- by exit and safety signs throughout the building
- within lift cars
- near fire-fighting equipment
- near each fire alarm call point.

Lighting units should be approximately 2.0 m off the floor — they need to be seen and, if positioned too high, could become obscured by smoke as smoke rises and accumulates below ceiling level.

Fire Certificates will specify which type of emergency lighting system is to be installed and the location of the light fittings and illuminated signs will be annotated on the plan of the premises which forms part of the Fire Certificate.

What is required in respect of fire detection equipment?

All workplaces must have arrangements for detecting fire.

During working hours, fires are often detected through observation or smell but this may not always cover areas which are remote or where workers are only sporadically present, e.g. plant rooms.

Fire detection systems do not need to be complex but they do need to be effective.

Fire detection systems can include:

- smoke detectors
- heat detectors
- flame detectors
- 'beam' detectors.

Fully integrated detection systems are becoming increasingly common and these are linked to fire alarm systems.

Control panels are sited in easily accessible positions and indicate where a detector has been activated, i.e. in what zone. Heat, smoke or flame may have been detected and the fire alarm activated.

Smoke detectors should be interlinked and should operate on both mains and battery power.

In smaller workplaces a series of interlinked smoke detectors will probably be all that is required for an effective 'early warning system'.

What is required in respect of fire alarm systems?

All work premises and premises to which the public resort must have a means by which the alarm can be raised in the event of a fire.

Raising the alarm gives people warning that a fire has broken out and encourages them to evacuate the building.

The fire warning system needs to be appropriate for the workplace but it must be effective and available for everyone to use.

In a smaller workplace where people work closely together, a manually operated klaxon could be used or a verbal alarm, e.g. shouting 'Fire! Fire!'.

Fire warning alarms must be heard in all parts of the building and part of the Fire Risk Assessment must cover the effectiveness of the fire alarm system.

Larger workplaces generally require a more sophisticated alarm system — often an electrically operated break glass call point system.

Manual call points are usually a break glass call point where the glass is broken and the 'button' pushed. This sets off the alarm sounders — often a continuous bell or siren.

A break glass call point will usually set an alarm off throughout the building and indicate evacuation. Occasionally, the fire alarm may be 'zoned' so that alarms go off in one area.

Integrated detection and alarm systems will rely on the identification of smoke or heat in an area and, when predetermined temperatures are reached or smoke is detected, the fire alarm will activate. Automatic fire alarms do not need someone to activate them.

However, every automatic detection and alarm system should be supplemented by a manual alarm call system.

All alarm call points should be on fire exit routes — people need to leave the area first and should raise the alarm on their way out of the building. Call points should also be on staircases and final exit routes.

A good practice guide is that no one should travel more than 30 m to reach a call point.

The audible fire alarm must be heard in all parts of the building and must be different to any other audible systems in use in the workplace.

Fire alarms should operate between 65 dB(A) and 75 dB(A). The latter noise level is essential if people will be sleeping on the premises.

It is common for the fire alarm system in places of public access (e.g. shops, shopping malls, entertainment complexes), to be incorporated into the public address system so that verbal instructions to evacuate the building are given. However, any public address system used as the fire alarm system must be capable of operating in a power failure so a back up, ancillary power source will be required.

The Fire Risk Assessment must consider what type of alarm is provided within the workplace and whether it is adequate.

The needs of all people must be considered and fire alarms may need to incorporate flashing lights so as to be seen or understood by everyone, e.g. people with learning disabilities, hearing impairment.

Visually impaired people need to know that a fire alarm has been activated. This is where a public address system may help.

What is a sprinkler system?

A sprinkler system is a fixed fire extinguishing system installed in many commercial buildings. They are normally used for general protection throughout the building and protect contents, fabric and structure from fire by drenching the fire with water.

Sprinkler systems are increasingly recognised as providing effective life safety for persons in a building and the number of fatalities in buildings in which sprinklers are installed is significantly less than for buildings without a sprinkler system.

The British Automatic Sprinkler Association describes a sprinkler system as follows:

> All areas of the building to be protected are covered by a grid of pipes with sprinkler heads fitted into them at regular intervals. Water from a tank via pumps or from the mains water supply (depending on mains pressure) fills the pipes.

Each sprinkler head has a sensor which will open when it reaches a specific temperature and spray water out and down onto the fire.

The heat and hot gases from the fire are sufficient to activate the sensors of the sprinklers.

Only sprinklers over the fire which have been activated will open to discharge water. Others remain closed until the temperature in the area rises, thereby triggering them.

The localised triggering of the sensor heads depending on temperature rise minimises the damage to areas not affected by the fire.

Sprinkler heads are spaced generally below the ceiling and designed so that if one operates there is still enough water to be supplied to the other heads.

Fire engineers need to calculate the size and length of pipes and the volume of water which the system needs to work effectively.

Sprinkler heads can be placed in closed roof voids, ducts, lift shafts, etc. and in any areas where a fire could start without being detected.

Sprinkler heads can be isolated by various valves, thus allowing for easier maintenance and repair.

What fire extinguishers will be required?

The Fire Risk Assessment should identify the type of fire extinguishers which need to be placed in different parts of the building as the Fire Risk Assessment should have identified which type of fire is more likely to occur in a given area.

If the premises are covered by a Fire Certificate, usually the Plan attached to the schedule will identify what type of fire extinguisher is required and where they should be placed.

All workplaces must be provided with suitable fire-fighting equipment for use by people in the premises.

When considering what type of extinguisher to provide, it is necessary to identify the type of material likely to burn.

What is meant by classification of fires?

Fires are classified in accordance with a British and European Standard as follows:

Class A: fires involving solid materials where combustion normally takes place with formation of glowing embers
Class B: fires involving liquids or liquefied solids
Class C: fires involving gases
Class D: fires involving metals
Class F: fires involving cooking oils or fats.

Class A and B fires

Class A fires usually involve solid materials, often of an organic matter such as wood or paper.

They can be dealt with using *water, foam*, or *multi-purpose powder* extinguishers.

Water and foam are the most suitable.

Class B fires involve liquids or liquefied solids such as paints, oils and fats.

These fires can be dealt with using *foam*, *carbon dioxide* or *dry powder*.

Carbon dioxide fire extinguishers are most suitable for use on fires in electrical equipment. (Note — there is no such fire as an 'electrical' fire.)

Class C fires

Class C fires are those involving gases.

Dry powder fire extinguishers are most suitable for use on these fires, but the most vital action to take is cutting off the gas supply. Consideration needs to be given to the possibility of explosion.

Class D fires

These fires involve metal and need to be dealt with by specialist fire fighters. Not only will the metals burn but chemical reactions could take place exacerbating the fire risks.

Class F fires

These fires involve cooking oils and fats and should be dealt with using a smothering medium such as a fire blanket. Some chemical extinguishing systems are designed into cooking canopies, e.g. Ansul system. These are specialist systems which should be operated under specialist conditions.

Is it still acceptable to have coloured fire extinguishers?

Yes, provided the fire extinguisher is in good working order and has been regularly maintained.

New fire extinguishers produced since 1997 now have to comply with the European Standard which requires the body or casing of all fire extinguishers to be red.

It is *not* a legal requirement to change all of the older style coloured fire extinguishers to the new red coloured cylinders. If an older type extinguisher needs to be replaced it will be changed for an all red coloured cylinder.

The new fire extinguishers can have a coloured band of approximately 5% of the external area on the front of the cylinder in the vicinity of the instructions for use.

Employers must consider whether there will be any risk of confusion by employees if they have a mixture of the new fire extinguishers (all red) and older style coloured fire extinguishers. It would be best practice to have one sort or the other. The Fire Risk Assessment should consider this aspect.

Fire extinguishers can have a life of up to 20 years so may not need to be changed.

Another way to reduce confusion over which fire extinguisher is which, is to display a wall-mounted sign indicating the type of fire extinguisher and the type of fire to use it on.

Where should fire extinguishers be sited?

Fire extinguishers should normally be sited in conspicuous locations on escape routes, preferably near exit doors.

Fire extinguishers should be grouped together to form a 'Fire Point'.

Fire points must be easily identified and could be indicated on the Fire Plan.

Fire Certificates may indicate the location of fire extinguishers.

Any room or work area which has been identified as being at risk of a fire should be provided with fire-fighting equipment, e.g. print and photocopy rooms should be provided with a carbon dioxide and water or general purpose fire extinguisher.

If fire extinguishers are placed in areas where they are not clearly visible, they must be adequately signed with a suitable safety sign which complies with the Health and Safety (Safety Signs and Signals) Regulations 1996.

Do fire extinguishers have to be wall hung?

Where practicable, fire extinguishers should be placed on wall brackets. This ensures that they are visible, not easily moved and being placed at height makes them easier to lift.

Heavier water fire extinguishers (or powder or foam) should be placed approximately 1 metre from the floor, with the handle easily accessible. This will ease the lifting process.

Smaller fire extinguishers (e.g. CO_2) could be placed at slightly higher positions.

If fire extinguishers cannot be wall-mounted they should be placed in conspicuous positions on a *base plate* on the floor. There are proprietary base plate or stands for fire extinguishers. Again, this should prevent them being moved.

When is it necessary to have a hose reel?

Usually, premises requiring a Fire Certificate may need a fire hose installed. Often they are provided in multi-occupied, multi-storey buildings which could fall into the high risk category.

Hose reels are common in hotels, halls of residence, etc.

A hose reel is a length of tubing with a shut off valve connected to a permanent pressurised water supply.

The Fire Risk Assessment would identify whether a fire hose is necessary. If in any doubt, consult the Fire Authority.

Fire hoses can sometimes lull people into a false sense of security about fighting a fire and they may be tempted to take risks. Fire hoses should only be used by trained, competent people.

Often, the Fire Brigade will use the fire hose as a quicker method of fighting the fire than dragging their own hoses up several flights of stairs.

When is it necessary to have a fire blanket?

Fire blankets are used to smother flames, usually involving cooking oils and grills, chip pans and fryers.

A fire blanket should always be provided in a kitchen. It must be easily accessible and should not be placed behind the equipment it is intended to be used on as the flames may prevent it being accessed.

Fire blankets can also be used in industrial process areas where fires need to be smothered and where molten metals are involved.

Fire blankets are classified as light-duty or heavy-duty.

Light-duty fire blankets are usually suitable for dealing with small fires in containers, e.g. chip pans.

Fire blankets can also be used to wrap around burning clothing — the flames need to be smothered quickly.

References

Fire Precautions Act 1971.

Fire Precautions (Workplace) Regulations 1997 (amended).

British Standard (BS) 5588: Fire Precautions in Buildings.

British Standard (BS) 5839: Fire Detection and Alarm Systems.

Case studies

October 2000 – Mexico City

At least 20 people killed and 27 injured in a nightclub. Survivors said they were blocked from leaving the burning building by disco personnel who insisted they pay their bills first.

Patrons panicked when smoke started filling the dance hall and began to scramble to escape out of the Club's only exit.

The building had a capacity for more than 1000 people, but had *no* emergency exit routes.

December 2000 – China

A fire started in the dance hall of a multi-storey commercial building, trapping construction workers on the second and third floors and over 200 people in the dance hall. The fire took 3 hours to put out and over 300 people died.

January 2001 – Netherlands

A fire started in the bar/café complex where 700 people were celebrating the New Year. The fire swept through the café, killing 10 people and injuring over 130.

Many of the injured were hurt by trampling over each other to smash windows so as to leap from the third floor to escape flames and smoke.

The cause of the fire is still unknown but may have been smuggled fireworks or an electric short circuit on the Christmas tree lights.

Only one of the three emergency exit routes was available for use.

February 2003 — Rhode Island

A fire broke out in a nightclub when pyrotechnics used as part of the band's act set fire to surrounding furnishing and ceilings, especially the polystyrene sound insulation material.

100 people died and many more were hospitalised as the fire swept through the building rapidly. Panic set in once the audience realised that the fire was not part of the pyrotechnic routine.

The fireworks ignited the foam causing a fast moving fire that consumed the club within minutes. Toxic fumes were given off by the foam.

6

Fire prevention

What is meant by the term 'fire prevention'?

Fire prevention usually refers to the practical measures and other procedures aimed at preventing outbreaks of fire and reducing the spread of flames, smoke and toxic fumes.

Fire prevention applies the common sense steps of preventing fires — after all, why accept a potentially fatal situation if you do not have to?

Preventing fires should be an essential part of any health and safety risk management policy.

What are the main causes of fire?

Usually fires start as a consequence of people's actions or lack of actions. Developing a 'fire safe aware' culture will significantly reduce risks of fire and will be a major part of the Fire Prevention Strategy.

The common causes of fire are:

- arson
- careless disposal of smoking materials

- combustible material left too near a heat source
- accumulations of easily combustible refuse or paper
- poor fire safety practices by contractors
- poor electrical practices, faulty electrical equipment, overloading of sockets, etc.
- obstructing ventilation grills to equipment causing overheating
- inadequate cleaning — especially of machines generating grease
- inadequate supervision and procedures regarding cooking activities.

Fires are often caused by a combination of factors and employers need to be aware of all possibilities.

Is it true to say that fires are inevitable?

There is a common misconception among many people that fires are 'inevitable', and that the only concerns an employer should have are regarding fire precautions and protection, i.e. minimising the effect of the fire.

This common perception was confirmed in the Public Enquiry report on the King's Cross fire, chaired by Mr Desmond Fennell QC when he said of London Underground Ltd:

> the management remained of the view that fires were inevitable on the oldest and most extensive underground system in the world. In my view, they were fundamentally in error in their approach.

The fire at King's Cross could have been prevented if better control of contractors had taken place, a prohibition of flammable substances was in place and better cleaning and degreasing of the escalator machinery had taken place.

The Bradford fire tragedy could have been prevented if fire prevention measures had been implemented regarding better housekeeping and the removal of combustible litter from under the wooden seating and stand.

Fires are not inevitable. Proper fire prevention policies will reduce or eliminate the likelihood of a fire starting and those businesses that recognise this pro-active risk management approach benefit greatly from reduced risk, reduced insurance premiums, etc.

A fire needs three components:

- heat/ignition source
- fuel
- oxygen.

Eliminating one or more of the three components will prevent a fire from starting.

What steps can be taken to prevent arson?

Arson — or malicious fire starting — is one of the commonest causes of commercial fires.

Usually, fires started by arson are more serious and cause greater damage than naturally occurring fires. This is because fires started intentionally are planned — the perpetrator usually knows that there is a good source of combustible material, the fuel source/ignition source is rapid (e.g. by the use of petrol or paraffin), and as there is usually no one in the building, there is no quick response to raising the alarm or fire fighting.

So, sensible steps to prevent arson involve reviewing security arrangements for the building as a first priority.

Preventing people from gaining access to the premises uninvited is essential — many arson fires are caused by youngsters who find ways of scaling fencing, etc. as a demonstration of their 'macho' prowess.

It will be necessary to conduct a Risk Assessment with the view of identifying the risks of an arson attack.

A systematic approach is needed to review the following:

- How easy is it to access the routes into the building — especially out of hours?
- Rear access routes — can people be seen easily — are there CCTV cameras, mirrors, etc.?
- External fencing and boundary walls — are they easily accessible — could extra precautions be taken, e.g. raising the height and repairing defects?
- Are windows left open or are they easy to open?
- Is the building left secure at night?
- If someone were to get in, would an intruder alarm sound? Could they be frightened off?
- Are there accumulations of combustible materials?
- Are flammable substances left easily accessible?
- Are matches, ignition sources, etc. easily found and useable?
- Is the building in good repair?
- Have all keys been accounted for?
- Have letter boxes or 'delivery chutes' which give access to the building been protected or could burning rags be pushed through gaps, etc.?
- Are doors and windows well fitting or are there gaps?
- Can access to your building be gained from adjacent premises?

In addition, the following need to be considered:

- control of people once on the premises
- control of contractors
- ease of access to the building for former employees who may have a grudge
- vetting of new employees — they may have a record for arson attacks
- implementation of 'locking up' procedures
- establishing a system of end of shift/day checks.

It has been shown that good external lighting aids building security and helps to prevent unauthorised access as people are more likely to be seen. So, if there is no external lighting, install some.

What steps can be taken to reduce the cause of fires from electrical equipment?

The locations of fires of electrical origin can be divided into three groups as follows:

- the fixed, permanent electrical installation of the building
- temporary wiring and leads to portable electrical appliances
- electrical appliances.

Fixed electrical installations must be installed in accordance with the Institution of Electrical Engineers Regulations for Electrical Installations — known as the IEE Wiring Regulations. The Regulations themselves have no statutory force but they are used as indications of good practice in respect of compliance with the Electricity at Work Regulations 1989.

The Electricity Supply Regulations 1988 require electrical supplies to be provided to consumers only if the installation is safe.

Fires may start in electrical cabling and circuits due to the following:

- overloading of cables by currents that the cables are not designed to carry. The cables overheat and the life of the insulation is shortened. Equipment may be operating by demanding too much current
- short circuit of conductors due to mechanical damage to insulation — extreme heat will cause combustible insulation to burn
- leakage of current to earth due to failure of cable insulation

- loose connections which result in local overheating of components, cables or combustible materials
- arcs and sparks that result from cable faults
- overheating of cables due to the presence of material insulation; the insulation may then deteriorate and a fire can result.

The Wiring Regulation are concerned with preventing fires through the use of fuses, circuit breakers and other devices which prevent overloading due to over currents from short circuits.

Cable specifications and current tolerances are linked to the capacity of the insulation in respect of fire resistance.

Some of the commonest fire prevention steps to take in respect of electrical systems and appliances are detailed below:

- have fixed wiring installations checked by a competent person every 5 years and obtain an IEE Wiring Certificate of Compliance
- employ only qualified electricians to install new sockets and fixed power lines to equipment; ask about the loading on the mains electrical circuit
- carry out an inventory of what equipment is fixed into the electrical supply and how many portable appliances are used
- carry out portable appliance testing in accordance with the Electricity at Work Regulations and Codes of Practice/ Guidance
- do not allow 'live' working on electricity
- check fuses in appliances to ensure that they are of the correct rating
- ensure that there are no 'amateur repairs' to plugs which use tin foil or nails as substitute fuses
- use residual current devices or earth leakage circuit breakers.

References

Fire Precautions Act 1971.

Fire Precautions (Workplace) Regulations 1997 (amended).

Electricity at Work Regulations 1989.

Fire safety — an employer's guide: HSE Books, Home Office.

7

Fire safety management in workplaces

Fire Risk Assessments have been carried out. What else must an employer do in respect of fire safety in the workplace?

Fire Risk Assessments are only one part of the fire safety responsibilities which employers have for the safety of their employees.

In addition, employers must:

- provide and maintain such fire precautions as are necessary to safeguard those who use the workplace
- provide information, instruction and training to employees about the fire precautions
- nominate persons to undertake any special roles which are required under the Emergency Plan
- consult employees about the nomination of people to undertake special roles in respect of fire safety and about proposals for improving fire precautions
- inform other employers who share the workplace or have workplaces within the building about any of the significant fire risks identified which might affect the safety of their employees
- co-operate with other employers about any measures needed to reduce or control fire safety hazards and risks

- ensure that fire safety requirements are managed in any common parts and that the Fire Regulations are complied with. This may need to be undertaken by persons in control of the premises
- establish suitable means for contacting the emergency services and ensure that they can be called easily.

The Management of Health and Safety at Work Regulations 1999 requires employers to provide Emergency Plans which outline the procedures to be followed in emergencies. Fire would be one such emergency.

What should the Emergency Plan for fire safety cover?

The Emergency Plan must be in writing and readily accessible for those who might need it.

It should be site-specific to the workplace and detail the pre-planned procedures in place for use in the event of a fire.

The Emergency Plan will usually include:

- action on discovering a fire
- warning if there is a fire
- calling the Fire Brigade
- evacuation of the workplace
- procedures for those particularly at risk, e.g. people with disabilities
- power or process isolation
- liaison with emergency services
- identification of key escape routes
- fire-fighting equipment provided
- specification of responsibilities in the event of a fire
- training required.

Emergency Plans should be tested regularly.

If possible, involve the Local Fire Authority.

What are some of the management controls which could be put in place for different aspects of fire safety?

The Fire Risk Assessment will identify the control measures necessary to manage fire safety risks in the workplace.

Some of the main fire safety controls which an employer needs to manage are detailed below.

Emergency exit routes

Emergency exit routes are those which need to be clear and unobstructed to allow people to get out of the building quickly in an emergency.

They could be the same as those that people use to enter the building. However, depending on how big the building is and how far away people are from an exit, additional exits are often required.

Escape routes must lead to a place of safety.

Check that fire escape routes are clear and unobstructed at all times. Make sure that they are easily accessible. Check that the area into which the escape route leads is clear and not obstructed. Can doors open outwards or are they blocked by cars, bins, barrels, etc.?

Check that any emergency lighting works.

Fire doors

Fire doors hold back smoke and flames. They should always be kept closed. Many fire doors are self-closing — do *not* prop them open.

Fire doors usually have 30 minutes' fire protection — they will hold back the flames and heat for 30 minutes before beginning to burn, so they keep fire escape routes safe for a short period of time.

Fire doors also hold back smoke if they are sufficiently close-fitting.

Fire doors should open in the direction of travel and be easy to use — with either a push bar or push pad for easy opening.

Fire doors must *never* be locked. If you need to keep a door locked for security reasons it must have a break glass bolt or lock so that it can be opened in an emergency.

Having locked fire exit doors is an offence and could cause you to lose your licence.

The following items should not be stored on protected fire escape routes or on any route that provides an emergency exit from a building:

- portable heaters of any type
- heaters that have unprotected naked flames or radiant bars
- fixed heaters using a gas supply cylinder, where the cylinder is in the escape route
- oil-fuelled heaters or boilers
- cooking appliances
- upholstered furniture
- coat racks
- temporarily stored items, e.g. furniture, packaging or deliveries
- lighting using naked flames
- gas boilers, pipes and meters (except those meeting Building Regulation Standards)
- gaming or vending machines
- electrical equipment other than normal lighting and emergency lighting.

Fire safety signage

Signs help people to know what to do and where to go when they are unfamiliar with premises and locations.

All premises *must* have clearly visible fire safety signs that:

- indicate fire exit routes
- identify fire doors
- identify fire alarm call points
- identify fire-fighting equipment.

Fire signs must now include a pictogram and a directional arrow. Words or text signs are no longer legal. Pictograms are symbols that should be understood by everyone, no matter what language they speak.

Signs must be clearly visible so need to be of the correct size.

Some fire exit signs must be illuminated — usually those on the final exit doors. If you have a Fire Certificate it will indicate on the Plan which signs are to be illuminated.

Even though you do not need to install illuminated signs it is a good idea to do so as it aids emergency evacuation in poor light or darkness.

An alternative provision where *no* illuminated emergency signs have been stipulated by the Fire Authority, is to display photo-luminescent signs. These glow in the dark and are much more easily seen.

Fire signs must be of the correct colour as follows:

Fire exit signs:	Green background White writing
Assembly point:	Green background White writing
Fire doors — keep shut or locked:	Blue background White writing
Fire call points: Fire extinguishers: }	Red background White writing

What should I put where?

A general sign to indicate a fire exit — customer areas, staff areas, and all other areas.

A fire exit which leads downwards, or straight on — suitable for all areas.

The fire exit is to the left of the sign — suitable for all areas.

The fire exit is to the right of the sign — suitable for all areas.

General fire exit sign — can be displayed above main entrance doors.

General fire exit which is to the left of the sign — suitable for all areas.

General fire exit which is to the right of the sign — suitable for all areas.

The fire exit is to the left of the sign — suitable for all areas.

Indicates that the fire door has a push bar to open and that it is a safe area to be using.

The fire exit is to the right of the sign — suitable for all areas.

Add this sign if the fire exit route is at an angle and leads upwards, e.g. upstairs from a basement.

Add this sign to a 'word' sign to make the word sign legal. Fire exit is to the left.

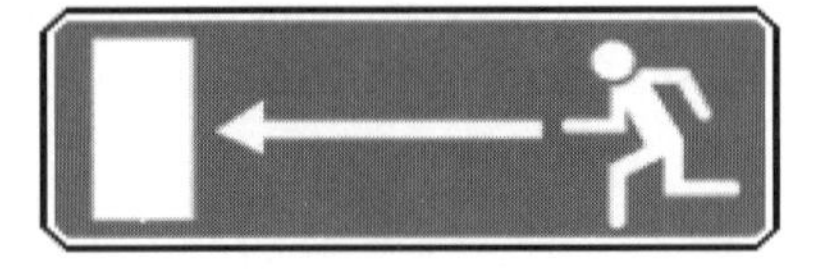

Exit towards the left for the nearest fire exit door.

Go downstairs to the nearest fire exit route.

Exit towards the right for the nearest fire exit door.

The ‘running man’ symbols which can be displayed next to text or word signs to make them legal.

Text and pictogram general fire exit signs, indicating whether to go to the right or left.

A sign which is usually displayed next to the fire extinguishers or where the fire alarm is located.

A sign which indicates that the fire alarm call point is nearby.

Usually displayed adjacent to water or some foam fire extinguishers.

Displayed next to carbon dioxide fire extinguishers.

A general sign displayed next to fire extinguishers.

A general fire extinguisher sign.

A fire alarm call point sign.

A fire hose reel sign, usually displayed next to the fire hose.

A telephone kept especially for raising the fire alarm.

Any of the following signs must be on doors which must be kept shut or locked for fire safety reasons e.g.:

- electrical cupboards
- lift motor rooms
- chemical storage cupboards
- all fire exit doors where they have self-closing devices.

Remember: fire exit route doors must *never* be kept locked shut.

Fire action signs

Signs displaying the actions you want people to take in case of a fire need to be displayed in suitable locations around the premises. Usually, they are displayed near fire call points and adjacent to emergency exits and/or fire extinguishers.

Fire action signs are simple and important pieces of information need to be added.

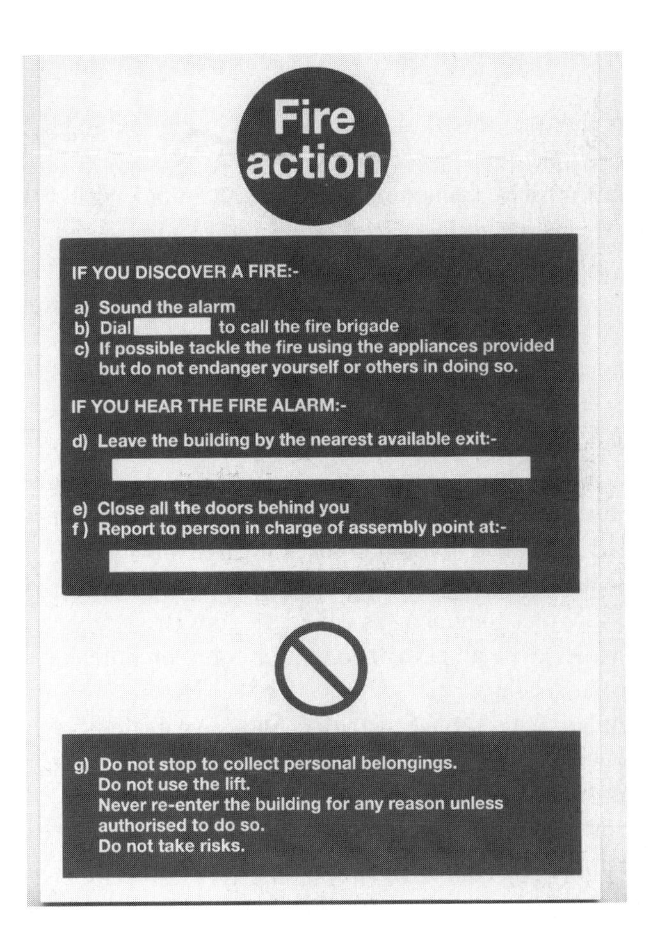

Assembly point

You need to designate an area where staff and others can report that they are safe. This could be the car park, if you have one, or a building across the road. Someone needs to be 'in charge' so that when the Fire Brigade arrives, they can tell them whether anyone is left inside the building or missing.

Evacuation procedures

If the fire alarm is raised or there is a fire in the building, you *must* evacuate people quickly and safely.

All staff must be trained in what to do in an emergency, i.e. who goes to check the toilets, work areas, store rooms and customer areas.

Choose someone to be responsible and to take charge of an incident. This person is often called the Fire Warden.

People must be asked to leave calmly and without panic.

Do not wait until you have checked for a fire or seen how 'real' the emergency is — leave at once.

Guide people to the emergency exit routes. Make sure that all areas of the building are checked — do not forget plant rooms.

All staff should be accounted for at the assembly point.

Do not use any lifts.

Make sure that someone has called the Fire Brigade.

If possible, close all doors and windows but do not take any risks with anyone's safety.

Remember that disabled customers may have difficulty getting out quickly — pay particular attention to their needs.

Fire drills are a good exercise for training purposes although difficult to carry out with customers present. Try to run through the fire drill and evacuation procedure regularly with staff — there are often opportunities to have a fire drill before opening times. Keep a record of when fire drills took place and who took part.

Top tips

Emergency exit routes

- Keep fire exit routes clear of obstructions at all times.
- Check them regularly throughout the day.
- Keep fire doors shut.
- Check that door seals are intact.
- Smoke is a greater killer than flames, so make sure that smoke cannot travel around the building.
- *Never* lock fire doors — use a break glass bolt if you need security.
- Obstructed fire exit routes can cause you to be closed down.
- You do not get a second chance to put it right — get it right first time.

Fire safety signage

- Designate an assembly point.
- Nominate someone to be in charge of an incident.
- Check that your fire signs can be clearly seen.
- Buy photo-luminescent signs as these show up clearly in the dark.
- Make sure that the details on the Fire Action Notice (see above) are completed and that these are displayed at exit routes and at fire call points.

Evacuation procedures

- Have practice fire drills.

- Make sure that staff know what to do to evacuate the premises.
- Nominate someone to check all areas of the premises, e.g. plant rooms.
- Keep emergency routes free of obstructions and clearly visible.
- Choose an assembly point and check staff off against the register.
- Do not forget other contractors and workmen who may be working in the premises.
- Appoint a 'Fire Warden' — someone to take extra responsibility for making sure that you and your staff are 'fire aware' and someone who takes charge of any incident.

How often should I carry out a fire safety check in my premises?

Regular fire safety checks reduce the risks from fire, prevent complacency and identify shortcomings in any of the control procedures in place.

The Management of Health and Safety at Work Regulations 1999 require employers to control, monitor and review any arrangements or preventative measures implemented in the workplace in order to control hazards and risks.

Fire safety controls can easily be abused, e.g. fire exit routes become blocked, fire extinguishers may be empty, the fire alarm may not work.

A systematic and regular procedure for fire safety checks will help to prevent deterioration in control measures.

The Fire Risk Assessment will identify the control measures necessary to reduce or eliminate the hazards from fire and during the risk assessment process consideration should be given to how frequently those controls need to be checked.

Fire safety checks should become second nature and a quick visual check of the workplace should be carried out every day before work starts.

If employees have had training in fire safety they will be familiar with recognising hazards and should be encouraged to play an active part in managing fire safety in the workplace.

A visual check of fire exit routes every day will soon identify whether they are obstructed. If they are, steps can be taken then and there to remedy the defect.

More detailed fire safety checks could be undertaken monthly by using a fire safety checklist and recording the findings.

As with any control system for health and safety, the employer will need to show that they were aware of the hazards and risks and had an effective control procedure in place. An 'effective control procedure' can only be demonstrated if it is regularly reviewed and shown to be working.

What subjects/areas should be covered in a fire safety check?

A fire safety check should be an objective review of the fire safety hazards, risks and precautions associated with the workplace.

The fire safety check should be relevant to the hazards and risks associated with the workplace and should not be a long list of every eventuality in respect of fire risks.

It is probably best practice to develop individual company fire safety checklists. There are examples available in various publications but it should not take too long to devise a 'bespoke' form.

A fire safety checklist should include the following.

- Good housekeeping
  - are areas free of rubbish?
  - is rubbish removed frequently?
  - ensure no excess or combustible materials
  - personal workplaces to be kept tidy.
- Storage areas
  - identify what is stored where
  - ensure that refuse is removed
  - ensure safe storage of items
  - make room to move around piles of stock
  - check that there is no stock obstructing vents, etc.
  - no stock to obstruct sprinkler heads or fire alarm systems
  - storage areas to be kept separate from work areas — doors kept shut, etc.
- Smoking
  - is a no-smoking policy in place
  - are smokers' materials dealt with safely, e.g. ashtrays emptied into metal containers
  - are smoking rooms checked frequently
  - are fire detection devices working?
- Flammable substances
  - are any stored in the work area
  - are they needed

  - are they kept away from ignition sources
  - are they used by trained people
  - has consideration been given to fumes and vapours
  - are ventilation grills unobstructed
  - are metal, non-combustible storage boxes used
  - is a 'with permit' procedure in operation
  - are lids kept on containers, etc.?
- Equipment maintenance
  - is routine maintenance carried out?
  - ensure no 'friction' burns
  - machinery to be kept clean
  - no build-up of grease to be allowed
  - no oily rags to be kept near equipment
  - no combustible materials to be kept near by.
- Flammable gases
  - any use of LPG
  - are gases stored externally
  - are they kept in a separated area
  - is a 'with permit' system in use
  - empty cylinders to be stored safely
  - are regular checks carried out?
- Heating and lighting
  - are heating appliances correctly used and positioned?
  - no use of employees' own heaters to be permitted
  - regular maintenance to be carried out
  - no combustible materials allowed in the vicinity
  - correct light bulb wattages to be used
  - no combustible lamp shades permitted
  - nothing to be stored under lights — be aware of radiant heat.
- Electrical appliances
  - are they used correctly
  - tested regularly
  - maintained properly
  - are visual checks carried out daily
  - are correct fuses, plugs, etc. used

  - are safety shut off switches accessible
  - are all safety devices working?
- Arson
  - are good security systems in place
  - are regular perimeter checks carried out
  - are combustible materials locked away
  - ensure that no ignition sources are available?
- Fire protection systems
  - are fire alarms working
  - is fire-fighting equipment available
  - is all equipment regularly checked
  - have sprinklers been checked
  - are smoke and heat detectors in place
  - have you ensured that there are no obstructions?
- Means of escape
  - are these clear and accessible from all areas
  - are all routes unobstructed
  - are doors kept shut
  - are panic bolts working
  - ensure that there are no locked fire doors
  - is suitable lighting installed
  - is emergency lighting working
  - is signage displayed clearly
  - are all instructions unambiguous
  - are assembly points clearly identified
  - are fire action notices displayed?
- Staff training
  - has everyone had fire safety induction training
  - have fire wardens been appointed and trained
  - are contractors given appropriate information on fire procedures
  - are visitors given information?
- Emergency procedures
  - who does what and when
  - raising the alarm
  - calling the Fire Brigade

  - meeting the Fire Brigade
  - searching the building
  - fighting the fire.
- Management procedures
  - completion of Fire Risk Assessments
  - review of Risk Assessments
  - availability of emergency plan
  - appointment of competent persons
  - monitoring and review processes
  - co-ordination of departments with regard to fire hazards
  - use of 'permit to work' and 'Hot Works Permit' systems
  - control of contractors
  - staff training programme.

The above points are some of the key subjects which can form the basis of a fire safety checklist. The format can be questions and yes/no answers, or statements and satisfactory/non-satisfactory answers. Any fire safety checklist, however, should contain a section which identifies any shortcomings in the control measures needed, with remedial action to be taken, by whom and by when. The findings of the fire safety check should be reviewed to ensure that remedial works have been completed.

FIRE SAFETY CHECKLIST

A regular fire safety check of the premises should be carried out. Complete this checklist monthly and keep a record. You should also make a daily pre-opening check that fire safety measures are satisfactory.

Fire Safety Management	**Yes**	**No**	**Not applicable**
1. Do you carry out daily checks on all fire safety measures?			
2. Are staff trained in emergency procedures?			
3. Have you got an Emergency Plan?			
4. Have you completed Fire Risk Assessments and/or updated earlier versions?			
5. Is the Fire Certificate available at all times?			
6. Do you comply with conditions of any special licenses?			
Means of escape			
7. Are all escape routes clear of obstructions?			
8. Are doors kept shut on escape routes?			
9. Are all internal fire doors clearly labelled?			
10. Can all fire safety signs and call point signs be seen clearly?			

Fire Safety Management	**Yes**	**No**	**Not applicable**
11. Are any fire doors locked or not able to be opened?			
12. Are corridors used for any storage of combustible materials, etc.?			
Fire alarm			
13. Is the fire alarm tested weekly and records kept?			
14. If a manual alarm is used, is it tested and easily available?			
Emergency lighting			
15. Is it tested regularly?			
16. Are records kept of tests?			
Other detection equipment			
17. Are regular tests carried out and records kept?			
Fire instructions			
18. Do all staff know what to do in the event of a fire?			
19. Are Fire Action Notices displayed?			
20. Is an assembly point designated?			
Housekeeping			
21. Is refuse removed regularly?			

Fire Safety Management	**Yes**	**No**	**Not applicable**
22. Is combustible material kept away from heat sources?			
23. Are aerosol cans stored safely?			
24. Are boiler rooms and electrical cupboards kept free from combustible materials?			
Electrical			
25. Are sockets overloaded?			
26. Is the 'one plug, one socket' rule followed?			
27. Have checks been carried out on electrical plugs, leads and appliances?			
28. Is equipment earthed?			
Smoking			
29. Are cigarette ends properly discarded into metal receptacles?			
30. Is furnishing non-combustible?			
31. Are staff trained on what to look out for with regard to smoking material and smouldering fires, etc.?			
32. Do staff follow 'no smoking' rules?			

Fire Safety Management	Yes	No	Not applicable
Heating			
33. If LPG heaters are used, are they serviced regularly?			
34. Are LPG heaters switched on by trained people?			
35. Are LPG cylinders stored outside the premises?			
36. Is combustible material kept away from heaters?			
37. If open fires are used are they monitored, with fire guards?			
38. Are heaters securely fixed?			
Lighting			
39. Are combustible materials, equipment, etc. kept away from light bulbs?			
40. Is the correct light bulb wattage used for the lampshade?			
Flammable substances			
41. Are substances kept away from heat sources and in metal containers?			
42. Are cleaning materials correctly stored away from heat?			

Fire Safety Management	**Yes**	**No**	**Not applicable**
43. Are chemicals used safely and not near to naked flames?			
Storage			
44. Are storage areas easy to get to, are electrical sockets visible and is stock away from any heat source?			
Kitchens			
45. Are canopies cleaned regularly?			
46. Are deep fat fryers thermostatically controlled?			
47. Are grease deposits cleaned regularly?			
48. If heat lamps/blow lamps are used for flambé dishes, is there a Risk Assessment?			
49. Are thermostats working and maintained?			
50. Are electrical and gas equipment serviced regularly?			
Night checks			
51. Is there a proper procedure for checking the premises at night?			

Fire Safety Management	Yes	No	Not applicable
Other			
52. Are contractor's Hot Works controlled?			
53. Are contractor's works checked?			
54. Are all employees fully trained in fire safety?			
55. Is there any activity in the vicinity which could cause an increased fire risk to the premises?			

Are there any guidelines which indicate how often different types of fire safety equipment needs to be checked?

Guidelines on the frequency of fire safety checks for equipment are contained in the publication *Fire safety — an employer's guide* issued by the Government.

In addition, the Loss Prevention Council and Fire Protection Association issue guidance on the testing and maintenance of fire safety equipment.

Manufacturers' instructions on maintenance must also always be followed.

The following provides a summary of weekly checks.

Fire detection, e.g. automatic detectors, smoke alarms, heat detectors	Weekly test to ensure that they are working. Weekly visual inspection.
Fire alarm systems	Weekly test to ensure that they are working. Weekly check of call points.
Emergency lighting	Weekly check of all manual systems, e.g. torches. Visual check of all light fittings.
Fire extinguishers	Weekly check to ensure all in the correct location. Weekly check to ensure that all are full.

Monthly checks of all fire safety equipment should:

- test that all equipment is working
- ensure that emergency call points operate
- test that all emergency lighting works
- test smoke and heat alarms
- ensure that fire-fighting equipment is in place.

Annual checks should be carried out by a competent person and in accordance with manufacturers' instructions. A full check and test of the entire system should be carried out by service engineers.

All checks, whether weekly, monthly or annual, must be recorded. The following record sheets are examples of the type of forms which can be used.

FIRE EXTINGUISHER AND HOSE REELS – RECORD OF TESTS AND INSPECTION

- Check all equipment for correct installation and function on a weekly basis.
- Arrange for annual servicing and testing by a competent engineer.

Date	Type	Location	Tests or inspection	Satisfactory Yes/No	Remedial action taken and signature

FIRE ALARM SYSTEM – RECORD OF TESTS

- Check for state of repair and operation on a weekly basis.
- Arrange for annual servicing and testing by a competent engineer.

Date	Fire alarm	Provides effective warning throughout	Remedial action taken and signature
	Call point use	Satisfactory Yes/No	

EMERGENCY LIGHTING SYSTEM — RECORD OF MAINTENANCE AND TESTS

- Check all lights for state of repair and correct functioning on a *weekly* basis.
- Arrange for annual servicing and testing of the system on an *annual* basis by a competent engineer.

Date	Satisfactory Yes/No	Type of defect and location	Remedial action taken and signature

FIRE DETECTION SYSTEM – RECORD OF TESTS

- Arrange for *annual* servicing and testing by a competent engineer.

Date	Type of detectors Smoke/ heat	Location	Detector activated	Satisfactory Yes/No	Action required	Signed

SUPPRESSION SYSTEMS – RECORD OF TESTS

- Arrange for annual servicing and testing by a competent engineer.
- Use this form to record tests on any sprinkler system or Ansul system, and any other fire suppression system installed (if any).

Date	Type of system	Location	Detector activated	Satisfactory Yes/No	Action required	Signed

Are there any structural features of a building that can be checked to ensure fire safety?

The most serious feature which increases the risk of fire spreading throughout a building is unprotected openings in compartment walls, floor and ceilings.

Fire safety design works on the principle of compartmentation which ensures that if a fire starts in an area it cannot spread beyond that area.

Any compromise to compartmentation will allow heat, fumes, smoke and toxic fumes to spread rapidly around a building.

Structural features to check are:

- ducts without dampers
- flues
- redundant chimneys
- voids behind panelling, floors and ceilings
- holes around service pipes
- uncompartmented roof voids
- unprotected staircases
- warped and ill-fitting doors
- missing or removed doors
- removed partition walls.

It may be necessary to consult architects' drawings, design specifications or the mechanical and electrical manuals and drawings.

Repairs and refurbishment works often create fire safety hazards by breaching compartmentation and the contractors often fail to reinstate the integrity of the wall, ceiling, etc.

When refurbishment works, new air conditioning, etc. are to be undertaken, it is imperative to agree remedial actions. An architect should advise, but for those projects undertaken in house, the Project Manager should have regard to the likely breach of structural fire safety measures.

What is a 'Hot Works Permit'?

A Hot Works Permit is a safety management procedure for controlling the use of equipment and processes which use heat-generating equipment and which, therefore, pose a risk of fire. It is a formal written system.

A Hot Works Permit is issued for every job which uses hot work/heat-generating equipment or heat-generating substances.

A Hot Works Permit records the safety management procedures in respect of certain job activities.

Any activity involving:

- blow torches
- hot bitumen boilers
- welding
- use of LPG
- use of flammable substances

should be accompanied by a Hot Works Permit.

What information should a Hot Works Permit contain?

The Permit to Work form must help communication between everyone involved in carrying out the task, or who will be affected by the task.

The main elements of a 'Permit' system are:

- permit title
- job location
- identification of plant to be worked on
- description of work to be done
- equipment to be used
- hazard identification
- control measures necessary
- protective equipment or clothing to be used

- training and competency of individuals
- identification of persons who will be carrying out the tasks
- authorisation by competent person
- handback procedure
- acceptance of safety procedures
- final checks on handback
- signatures
- site safety rules.

Each Permit should be designed for the specific job and should address the hazards and risks and outline the controls to be taken to eliminate or reduce risks.

Permits should be kept for a reasonable period of time and it may be necessary to review the safety precautions taken on a specific job.

References

Fire Precautions Act 1971.

Fire Precautions (Workplace) Regulations 1997 (amended).

Management of Health and Safety at Work Regulations 1999.

Health and Safety at Work Etc. Act 1974.

8

Hazardous and flammable substances

What are flammable liquids?

Flammable liquids give off large volumes of flammable vapours at room temperature.

These vapours, when mixed with air, can ignite, often violently.

Spilled flammable liquids can flow considerable distances to an ignition source and the source of ignition could be in a completely different area to the spilled liquid.

Spills of flammable liquids on clothing create considerable fire risks.

Flammable liquids are therefore considerable fire safety hazards and need to be closely managed and controlled so as to eliminate or reduce their contribution to a fire.

A flammable liquid may be reasonably safe in a stored *lidded* container but, if the lid is poorly fitting, fumes can escape and provide the potential for ignition.

What controls need to be implemented for flammable liquids?

The following guidelines should be followed when storing *any* flammable liquids.

(1) Why store flammable liquids?
(2) Substitute the product for something less flammable.
(3) Store flammable liquids in a separate storage area or in a purpose-made, fire-proofed bin or cupboard.
(4) Dispense and use them in a safe place, where there is good ventilation and no heat or ignition sources.
(5) Keep containers closed when not in use. Check lids regularly.
(6) Use safety containers with self-closing lids if possible.
(7) Dispense liquids over a tray or shallow container so as to contain any spillage to a localised area.
(8) Keep non-flammable, absorbent material close to hand to deal with spillages, e.g. sand.
(9) Clear up spillages immediately.
(10) Dispose of contaminated rags and materials safely and remember that these have become flammable substances. Dispose of separately and in flame-proof containers. Do not store quantities of contaminated materials together as they may 'spontaneously' combust due to heat generation and ignite the vapours given off from the materials.

What are flammable dusts?

Fine particles of dust in certain circumstances can accumulate and explode, causing both physical damage and potential fire risks.

Dusts from solid materials which are flammable or which contain flammable substances or components will, in themselves, be flammable and any accumulation has the potential to ignite.

The following controls in relation to dusts are required:

- keep dust quantities to a minimum and prevent escape to the atmosphere
- use local exhaust ventilation to remove dusts at source
- regularly empty dust bags so as to reduce the volumes collected
- keep machinery clean
- clean the workplace by regular vacuuming, etc.
- clear up spillages as they occur
- follow any inherent dust control procedures and safety procedures associated with the plant.

What are flammable solids?

Some types of 'solid' material will ignite easily and burst into flames with a fierce and hot burning intensity.

Substances such as plastic foam, packaging, polyester wadding and textiles may fall into this category.

Safety procedures to follow are detailed below:

- do not store flammable materials close to any heat source, e.g. heaters, fans, cookers, electrical equipment
- illuminate all potential sources of ignition
- store small quantities of materials
- do not use highly flammable materials in high risk fire areas
- keep all passageways, work areas, etc. clear so that if a fire were to start it could not 'leap' to adjoining fuel sources
- maintain good general fire precautions.

What are flammable gases?

Gases are often stored in cylinders under pressure and any uncontrolled release of the contents can cause explosion and fire.

Flammable gases will be those of the liquifuel petroleum gas family, i.e. calor gas and propane.

Other common flammable gases are methane, acetylene and butane.

Examples of flammable materials

Flammable liquids

- petroleum products
- petrol, diesel
- cooking oils
- motor oils
- hydraulic fluids
- lubricants
- solvents
- paints
- thinners
- degreasing agents

Flammable dusts

- flour
- confectionary dusts
- swarf
- wood dusts

Flammable substances/solids

- foam
- polystyrene
- textiles
- polyester wadding
- building products, e.g. insulation materials

Safety procedures are as follows:

- store cylinders of gas safely and securely
- store cylinders in an open air secure cage or compound
- secure all valves in the 'shut' position
- protect cylinders from impact damage
- use and fit the correct equipment, e.g. pressure valves, hoses and connectors
- check hoses, valves, etc. regularly
- do not use faulty equipment
- do not tamper with valves unless trained and competent to do so.

Why is oxygen a fire hazard?

In certain circumstances, known as an oxygen-enriched atmosphere, oxygen will fuel a fire rapidly and will contribute to the risk of explosion.

Materials that will ordinarily burn slowly may burn very vigorously in an oxygen-enriched atmosphere.

Some substances may bust into flames in an oxygen-enriched atmosphere, e.g. greases and oils.

Safety procedures to follow include the following:

- never use oxygen in place of compressed air
- never use oxygen to 'sweeten' the air in a confined space or working area, e.g. tanks or vats
- never use grease or oil on equipment containing oxygen.

What is an oxidising substance?

An oxidising substance is one which can cause a flammable material to ignite because it releases oxygen and increases the potential for combustion.

Oxidising substances must carry a hazard warning sign and must not be stored adjacent to flammable materials.

As an employer, I need to store flammable substances/liquids in my workplace. What safety measures do I need to take?

The measures stated in the previous questions must be followed, depending on what type of flammable substance you have.

In addition, as an employer you must:

- make sure that workers know about the hazards and risks of these substances
- ensure that employees are aware of the Risk Assessments
- ensure that employees have had training on how to use flammable substances, how to store them safely, etc.
- ensure that employees are aware of the dangers of mixing chemicals, liquids and substances
- introduce Permit to Work and Hot Works Permit procedures
- introduce and monitor 'safe systems of work'
- train all employees in what to do in an emergency
- increase any fire-fighting provisions to address the increased risk
- apply for a special Fire Certificate if you store flammable substances, etc. above certain limits
- manage and maintain procedures for dealing with spillages, etc.
- consider the need for any special first aid equipment and specialist advice for treating injuries.

What are the Dangerous Substances and Explosive Atmospheres Regulations 2002 about?

The Dangerous Substances and Explosive Atmospheres Regulations 2002 (DSEAR) is a set of Regulations concerned with the protection against risks from fire, explosion and similar events arising from dangerous substances used or present within the workplace.

The Regulations apply to employers and the self-employed.

A dangerous substance is one which, because of its properties or how it is used, could cause harm to people from fires and explosions.

Dangerous substances include:

- petrol
- liquified petroleum gas
- paints
- varnishes
- solvents
- dusts which, when mixed with air, could cause an explosive atmosphere.

Dangerous substances are found in some quantity in most workplaces, e.g. paints, thinners, even if stored in the 'maintenance shed or office'.

An explosive atmosphere is an accumulation of gas, mist, dust or vapour which, mixed with air, has the potential to catch fire or explode.

What does an employer need to do under DSEAR 2002?

As an employer, you must:

- carry out a Risk Assessment of any work activity involving dangerous substances
- put in place measures to eliminate or reduce risks as far as is reasonably practicable
- provide equipment and procedures to deal with accidents and emergencies

- classify places where explosive atmospheres may occur in 'zones' and mark the zones as necessary
- provide information and training to employees.

The important part of the Regulations is to carry out a Risk Assessment of the use and storage of dangerous substances.

Any Fire Risk Assessment should consider the need to manage the risks from substances likely to contribute to the risks of fire.

Emergency evacuation procedures may need to be reviewed because the spread of fire may be rapid and it will be vital to ensure that people can evacuate the building quickly.

How will I know if a substance is a dangerous substance under the Dangerous Substances and Explosive Atmospheres Regulations 2002?

The first step is to check whether the substances have been classified under the Chemicals (Hazard Information and Packaging for Supply) Regulations 2002 as:

- explosive
- oxidising
- extremely flammable
- highly flammable
- flammable.

The substance must carry one of the hazard warning symbols — an orange square with a black symbol.

Substances must be provided with Safety Data Sheets. These contain all the chemical, physical, hazard and safety information regarding the product and must be considered when determining whether substances fall within the remit of the Regulations.

Some substances may not carry a hazard warning sign but the way in which they are used in the work environment may cause them to be dangerous or highly explosive.

It is a combination of the substance and the circumstances in which it is used which is important.

Diesel oils, for example, are not classified as flammable under CHIP Regulations, yet if they are heated to a sufficiently high temperature they can create a fire risk.

References

Fire Precautions (Workplace) Regulations 1997 (amended).

Control of Substances Hazardous to Health Regulations 2002.

Dangerous Substances and Explosive Atmospheres Regulations 2002.

9

Fire safety on construction sites

What fire safety legislation applies to construction sites?

The following legislation may apply in some form or other to fire safety on construction sites:

- Fire Precautions Act 1971
- Fire Precautions (Workplace) Regulations 1997 and 1999
- Construction (Design and Management) Regulations 1994
- Construction (Health, Safety and Welfare) Regulations 1996
- Fire Certificates (Special Premises) Regulations 1976
- Management of Health and Safety at Work Regulations 1999
- Dangerous Substances and Explosive Atmospheres Regulations 2002.

Construction sites are places of work and therefore the duties placed on employers to maintain their premises in a safe condition apply to sites.

Construction sites carry a high risk of fire as the nature of the work increases the hazards and risks which contribute to a fire starting.

What are the legal requirements for fire safety on construction sites?

Fire safety on construction sites is specifically dealt with under Regulations 18, 19, 20 and 21 of the Construction (Health, Safety and Welfare) Regulations 1996.

These Regulations govern general health and safety on construction sites but include specific requirements for fire safety as follows:

- prevention of risk from fire
- provision of emergency routes and exits
- preparation of emergency procedures in the event of fire
- provision of fire-fighting equipment, fire detectors and fire alarm systems
- instruction of every person on a construction site in the actions to take in the event of a fire, including the use of fire-fighting equipment.

If the construction site is purely a construction site and no other activity is taking place, then the requirements of the Construction (Health, Safety and Welfare) Regulations 1996 are enforced by the Health and Safety Executive.

If the construction project is taking place on a multi-occupied site which continues to be occupied by the Client or other employer, or which is open to the public, then the Fire Authority enforce the requirements of fire safety as outlined in Regulations 19–21 above.

What causes a fire to start?

A fire needs the following to start:

- fuel
- ignition
- oxygen.

The above three component parts are often referred to as the 'triangle of fire'.

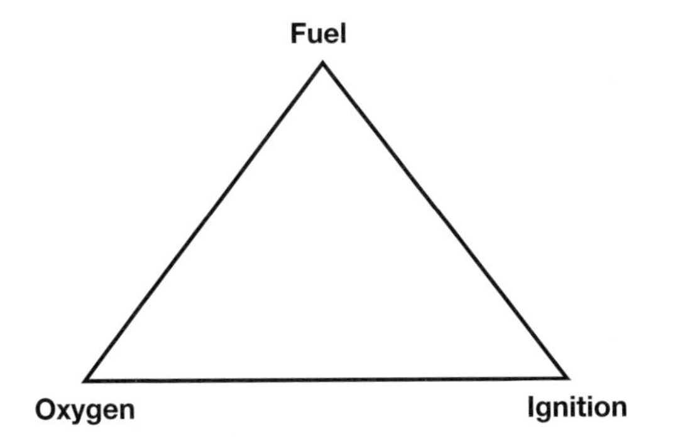

If one or more of the component parts of a fire are eliminated, a fire will not start or continue.

As the elements of fire are not successfully managed on all construction sites, the industry suffers approximately 4000 fires a year, costing millions of pounds in destroyed buildings, materials, plant and equipment and delayed projects.

Other insurance cover for buildings and construction work is difficult to obtain without a proper fire safety management plan in place.

As a Site Agent, what do I need to be able to demonstrate in respect of site fire safety management?

Generally, you need to demonstrate to HSE Inspectors and insurance assessors the following:

- recognition of the fire risks in the workplace and the extent of the risks
- assessment of the controls necessary to reduce the risks
- implementation of the controls necessary
- importance of constant review and monitoring of processes, controls, etc.

What is a Fire Safety Plan?

The Fire Safety Plan should identify fire risks throughout the site, e.g.:

- combustible materials
- use of hot flame equipment
- use of liquid petroleum gas
- use of combustible substances
- storage and use of any explosive materials and substances
- sources of ignition, e.g. smoking
- use of heaters.

Once the potential fire risks are identified, i.e. where, when, why and how a fire *could* start on site (or in the surrounding area, yards or outbuildings), the Fire Safety Plan should include precautions and procedures to be adopted to *reduce* the risks of fire. These could include:

- operating a Hot Works Permit system
- banning smoking on site in all areas other than the approved mess room
- controlling and authorising the use of combustible materials and substances
- providing non-combustible storage boxes for chemicals
- minimising the use of liquid petroleum gas and designating external storage areas
- controlling the location and use of heaters and drying equipment
- operating a Permit to Work system for all hazardous activities.

Having identified the potential risks and the ways to minimise them, there will always be some residual risk of fire. The Fire Safety Plan should then contain the Emergency Procedures for dealing with an outbreak of fire, namely:

- types and location of Fire Notices
- the location, number and type of fire extinguishers provided throughout the site

- the means of raising the alarm
- identification of fire exit routes from the site and surrounding areas
- access routes for emergency services
- procedure for raising the alarm
- assembly point or muster point.

The Fire Safety Plan should also contain the procedures to be carried out on site to protect against arson, e.g.:

- erection of high fencing or hoarding to prevent unauthorised entry
- fenced or caged storage areas for all materials, particularly those which are combustible
- site lighting, e.g. passive infra-red, security sensors
- use of CCTV
- continuous fire checks of the site, particularly at night if site security is used.

Procedures for the storage and disposal of waste need to be included as waste is one of the commonest sources of fire on construction sites.

Materials used for the construction of temporary buildings should be fire protected or non-combustible whenever possible, e.g. having 30-minute fire protection. The siting of temporary buildings must be considered early in the site planning stage as it is best to site them at least 10 m away from the building being constructed or renovated.

Having completed the Fire Safety Plan, a sketch plan of the building indicating fire points, assembly point, fire exit routes, emergency services access route to site, etc. should be completed and attached to the Plan. The sketch plan (which could be an architect's outline existing drawing) should be displayed at all fire points and main fire exit routes and must be included in any site rules/information handed out at induction training.

What is the most important part of fire safety on the construction site?

Fire safety needs to be managed on a construction site in the same way that hazards and risks associated with other activities need to be.

The same concept is applied to managing fire safety as to other site hazards: 'Fire Risk Assessment'.

The Fire Precautions (Workplace) Regulations 1997 (amended in 1999) apply to *all* employers and all workplaces and require employers to carry out Fire Risk Assessments and to record the significant findings in writing.

What is a Fire Risk Assessment?

A Fire Risk Assessment is a structured way of looking at the hazards and risks associated with fire and the products of fire, e.g. smoke.

Like all Risk Assessments, a Fire Risk Assessment follows *five key steps*, namely:

Step 1. Identify the hazards.
Step 2. Identify the people and the location of people at significant risk from a fire, i.e. who they are and where they work.
Step 3. Evaluate the risks — how severe will things be.
Step 4. Record findings and actions taken to reduce any fire risks.
Step 5. Keep Assessment under review.

So, a Fire Risk Assessment is a record that shows you have assessed the likelihood of a fire occurring in your workplace, identified who could be harmed and how, decided on what steps you need to take to reduce the likelihood of a fire (and therefore its harmful consequences) occurring. You have recorded all these findings regarding your construction site into a particular format, called a Risk Assessment.

How do I do a Fire Risk Assessment and how often?

A fire needs three things to start:

- oxygen
- fuel
- heat or ignition source.

If any one of these is missing, a fire cannot start.

Once a fire has started it spreads and engulfs other sources of fuel. Also, as a fire intensifies it gives off more heat. In turn, this heat causes things to ignite.

Step 1. Identify the hazards

Sources of ignition

You can identify the sources of ignition on your site by looking for possible sources of heat that could get hot enough to ignite the material in the vicinity.

Such sources of heat or ignition could be:

- smokers' materials
- naked flames, e.g. fires, blow lamps, etc.
- electrical, gas or oil-fired heaters
- Hot Work processes, e.g. welding and gas cutting
- cooking, especially frying in the mess room
- faulty or misused electrical appliances including plugs and extension leads
- lighting equipment, especially halogen lamps
- hot surfaces and obstructions of ventilation grills causing heat build-up
- poorly maintained equipment that causes friction or sparks
- static electricity
- arson.

Look out for any evidence that items have suffered scorching or overheating, e.g. burn marks, cigarette burns, scorch marks, etc.

Check each area of the site systematically:

- access routes
- work areas
- storage areas
- messing facilities
- welfare facilities
- fuel storage areas.

Sources of fuel

Generally, anything that burns is fuel for a fire. Fuel can also be invisible in the form of vapours, fumes, etc. given off from other less flammable materials.

Look for anything on the site that is in sufficient quantity to burn reasonably easily, or to cause a fire to spread to more easily combustible fuels.

Fuels to look out for are:

- wood, paper and cardboard
- flammable chemicals, e.g. cleaning materials
- flammable liquids, e.g. cleaning substances, liquid petroleum gas
- flammable liquids and solvents, e.g. white spirit, petrol, methylated spirit
- paints, varnishes, thinners, etc.
- furniture, fixtures and fittings
- textiles
- ceiling tiles and polystyrene products
- waste materials and general rubbish
- gases.

Consider also the construction of the building, if appropriate — are there any materials used which would burn more easily than other types? Hardboard, chipboard and blockboard burn more easily than plasterboard.

Identifying sources of oxygen

Oxygen is all round us in the air that we breathe. Sometimes, other sources of oxygen are present that accelerate the speed at which a fire ignites, e.g. oxygen cylinders for welding.

The more turbulent the air the more likely the spread of fire will be, e.g. opening doors brings a 'whoosh' of air into a room and the fire is fanned and intensifies. Mechanical ventilation also moves air around in greater volumes and more quickly.

Do not forget that while ventilation systems move oxygen around at greater volumes, they will also transport smoke and toxic fumes around the building.

Step 2. Identify who could be harmed

You need to identify who will be at risk from a fire and where they will be when a fire starts. The law requires you to ensure the safety of your staff and others, e.g. contractors. Would anyone be affected by a fire in an area that is isolated? Could everyone respond to an alarm, or evacuate?

Will you have people with disabilities on the site, e.g. visually or hearing impaired? Will they be at any greater risk of being harmed by a fire than other people?

Will contractors working in plant rooms, on the roof, etc. be adversely affected by a fire? Could they be trapped or fail to hear alarms?

Who might be affected by smoke travelling through the building or site? Smoke often contains toxic fumes.

Step 3. Evaluate the risks arising from the hazards and the control measures you have in place

What will happen if there is a fire? Does it matter whether it is a minor or major fire?

A fire is often likely to start on a construction site because:

- people are careless in discarding cigarettes and matches in case they get 'caught'
- people purposely set light to things
- Hot Works are not controlled
- people put combustible material near flames or ignition sources
- equipment is faulty because it is not maintained, e.g. electrical tools
- poor electrical safety is allowed to be the norm in site offices and mess rooms.

Will people die in a fire from:

- flames
- heat
- smoke
- toxic fumes?

Will people get trapped in the building?

Will people know that there is a fire and will they be able to get out?

Step 3 of the Risk Assessment is about looking at what *control measures* you have in place to help control the risk or reduce the risk of harm from a fire.

Remember — fire safety is about *life safety*. Get people out fast and protect their lives. Property is always replaceable.

You will need to record on your Fire Risk Assessment the fire precautions you have in place, i.e.:

- What emergency exits do you have and are they adequate and in the correct place?
- Are they easily identified, unobstructed, clear of boxes, furniture, etc.?
- Is fire-fighting equipment provided?

- How is the fire alarm raised?
- Where do people go when they leave the building — an assembly point?
- Are the signs for fire safety adequate?
- Who will check the building and take charge of an incident, i.e. do you have someone appointed to manage incidents?
- Are fire doors kept closed and unlocked?
- Are ignition sources controlled and fuel sources managed?
- Do you have procedures to manage contractors on the site, especially those using Hot Works?

Are operatives trained in what to do in an emergency? Can they use fire extinguishers? Do you have fire drills? Is equipment serviced and checked, e.g. emergency lights, fire alarm bells, etc.?

Step 4. Record findings and actions taken

Complete the Fire Risk Assessment form and keep it safe.

Make sure that you share the information with all site operatives.

If contractors come to site, make sure that you discuss *their* fire safety plans with them and that you tell them what your fire precaution procedures are.

Step 5. Keep Assessment under review

A Fire Risk Assessment needs to be reviewed regularly — approximately every week and whenever something has changed.

Top tips

- Complete Fire Risk Assessments.
- Check fire evacuation procedures.
- Check all fire precautions identified as being necessary are in place and working.
- Ensure that operatives are trained. Share information with them. Fire safety is *everyone's* responsibility.
- Report any maintenance defects or carry out repairs — in a fire, in thick smoke and without lights, you do not want to trip over and break a leg.
- Never take risks with fire — discretion is the better part of valour!
- Vet contractors to make sure that they will work safely.
- Remember to consider arson.

What are general fire precautions on a construction site?

The term 'general fire precautions' is used to describe the structural features and procedures needed to achieve the overall aim of fire safety, which is to ensure that:

'Everyone reaches safety if there is a fire'.

Putting a fire out is secondary to 'life safety'.

Fire Risk Assessments are about ensuring that you have considered the likelihood of a fire starting, who it would affect, how quickly and how those people would be evacuated to a place of safety.

General fire precautions cover:

- escape routes and fire exits
- fire-fighting equipment
- raising the alarm
- making emergency plans
- limiting the spread of fire (compartmentation).

General fire precautions will invariably differ from site to site depending on the complexity of the site.

Life safety for all persons on the site must be properly considered, planned, implemented and regularly checked. Escape routes, for instance, should be permanent, clearly identified, lead to a place of safety, well sign-posted and unobstructed. An ad hoc scramble down ladders or jumping off floors will not be acceptable.

What are some of the basic fire safety measures?

The Fire Safety Plan should be developed as an integral part of the Construction Phase Health and Safety Plan or overall work plan if CDM Regulations do not apply.

The very basic requirements are:

- fire exit routes from the site
- fire-fighting equipment
- methods of raising the alarm
- lighting
- signage
- fire protection measures to prevent spread of fire
- safe systems of work
- prohibiting smoking on site
- appointment of Fire Wardens
- regular checks
- emergency procedures.

Fire exit routes

Where possible, there should always be more than one exit route from a place of work. If the travel distance is more than 45 m to an exit, there must be two or more exits. This travel distance will vary depending on the risk rating of the site.

If the number of exits cannot be improved, then the risk rating of the site for fire must be reduced.

Fire exit routes must be unobstructed, clearly defined, of adequate size and width and not locked.

Doors leading onto fire exit routes should open via a push bar in the direction of travel.

No fire exit route shall lead back into the building or site.

Fire exit routes must lead to a place of safety.

Fire exit routes must in themselves be protected from fire by fire protected enclosures or doors. Doors must be kept shut.

Exit routes using ladders, e.g. on scaffolding, need to be specially assessed as part of the site-specific Risk Assessment.

Fire exit routes must be clearly visible from all parts of the work area. Exit signs which meet the Health and Safety (Safety Signs and Signals) Regulations 1996 must be displayed.

If lighting is poor — use photo-luminescent signs.

All fire signage must display pictograms as a minimum. Text can also be used alongside the pictogram, as can directional arrows.

Fire-fighting equipment

Suitable fire extinguishers need to be placed in appropriate locations around the site, and always at fire points near the fire exit routes.

Multi-purpose foam or powder extinguishers are suitable but, so too, would be water and carbon dioxide. The Fire Risk Assessment should determine which type is required.

Fire-fighting equipment should be visible, properly signed, inspected weekly and ready to use if needed. Operatives should not need to climb over materials, move plant, etc. to use the extinguishers.

Either a designated number of operatives in each work area or all operatives should be trained in how to use the fire-fighting equipment.

Regular reassessment of the working area is needed to ensure that the locations of the fire extinguisher points are appropriate.

Methods of raising the alarm

A fully integrated alarm system would be beneficial on all sites, activated by break glass points and linked to an alarm control panel in the site office.

However, this is not always possible and alternatives are permissible, e.g.:

- hand bells
- klaxons
- sirens
- hooters.

The alarm in use on the site should be clearly identified and all operatives *must* receive training in fire alarm procedures.

Fire alarm points must be clearly visible, easily accessible, etc.

If alarms cannot be heard in all areas of the site there must be a procedure for Fire Wardens to warn Fire Wardens on other floors, etc.

On small sites, a simple shouting of 'Fire! Fire!' may be all that is needed.

If the site is multi-occupied without employers (e.g. a major department store refurbishment), then the construction site alarm system must integrate with that of other employers so that total building evacuation is occasioned as necessary.

Lighting

Emergency lighting is not necessarily required on all construction sites but if there is a risk of power failure and no natural daylight to the areas of work, emergency lighting will be essential.

A simple system of torches may suffice.

All emergency exit routes must be adequately lit at all times.

Emergency back-up lighting will activate when the main power supply fails.

Regular checks of emergency lighting will be necessary.

Signage

Signage enables people to be guided to safe places — either to emergency exit routes or safe places, e.g. to refuges or to assembly points.

Signs should be photo-luminescent, where possible.

Signs must be visible from all work areas, not confusing, of large enough size and accurate in the information they portray, e.g. they must not lead to a dead end as a fire exit route.

Fire protection measures to prevent fire spread

Generally, it is best to try to consider floors and different areas as compartments, with fire protected, closed doors, and fire protection to voids and ducts, etc.

Fire and smoke, including toxic fumes, spread rapidly. Compartmentation constrains them to one area.

Safe systems of work

Any work activity which has the potential to increase the risk of a fire starting *must* be controlled by a Permit to Work or Hot Works Permit system.

Hot Works should be prevented whenever possible. Controls need to be localised, e.g. additional fire extinguishers, regular checks, additional Fire Wardens.

Combustible materials, flammable gases, etc. should be removed. Flashover should be considered.

Only trained operatives should carry out Hot Works or use flammable materials, etc.

Smoking on site

There is no acceptable fire safety procedure other than to ban it completely.

However, smoking may be permitted in mess rooms. If so, strict controls must be implemented.

Appointment of Fire Wardens

Each floor or work area should have an appointed Fire Warden, i.e. people who are trained to know what to do in the event of a fire and how to evacuate their area, raise the alarm, etc.

There should be sufficient Fire Wardens to cover for absences. Fire Wardens should receive regular training.

Regular checks

Daily and weekly fire safety checks are advisable on all sites. Checks are *always* necessary after Hot Works, and usually approximately 1 hour after the end of Hot Works so that any smouldering materials can be identified.

Records of fire safety checks should be kept for the duration of the project. Remember — you need to demonstrate that you know what you are doing.

Emergency procedures

These must be specific for each site and written down clearly. Emergency procedures must be displayed in prominent positions.

They should include:

- type of fire alarm
- how to raise the alarm
- how to evacuate the site
- the assembly point
- the names of the Fire Wardens
- any highly hazardous areas
- storage of flammable materials
- procedures for visitors to site
- name and telephone numbers of local emergency services
- liaison with emergency services when they arrive on site.

Top tips

- Plan fire safety before works start.
- Reduce combustible materials.
- Reduce ignition sources.
- Keep fire exit routes clear.
- Display adequate and suitable fire signage.
- Train operatives in emergency procedures.
- Keep fire extinguishers on site, in suitable locations and of the correct type.
- Install emergency lighting if possible.
- Clearly describe the fire alarm raising procedure.
- Remember — get everyone out rather than fight the fire.

References

Management of Health and Safety at Work Regulations 1999.

Fire Precautions (Workplace) Regulations 1997 and 1999.

Fire Safety in Construction Work: HSG 168.

Fire safety — an employer's guide: HSE Books, Home Office.

Fire prevention on construction sites: Fire Protection Association, Fifth Edition, January 2000.

10

Fire safety in special and other premises

What fire safety requirements apply to places of public entertainment?

Premises used by the public must be as safe as is practicable from the risks of fire.

Many premises used for public entertainment need to be licensed by the Local Authority or the Licensing Justices and the conditions attached to the Licence will include provisions for fire safety and fire precautions.

Primary duties of fire safety in places of public entertainment fall to the Premises Manager. The Manager's duty is to protect the public by having a fully considered fire safety and emergency evacuation plan.

Places of public entertainment include theatres, concert halls, dance halls, discos, conference centres and indoor leisure centres.

Many public houses have Public Entertainment Licences, i.e. a licence issued by the Local Authority which allows them to have music and dancing on the premises. The PEL will stipulate the precautions to be taken in respect of fire safety.

Premises may need a Fire Certificate under the Fire Precautions Act 1971.

What does 'maximum occupancy number' mean?

The maximum occupancy number is the number of people who can occupy the building with a degree of safety because the number of escape routes and exits have been calculated to accommodate that maximum number.

Maximum occupancy rates are laid down either in the Fire Certificate or in the Public Entertainment Licence or in the licence granted by the Licensing Justices.

Maximum occupancy numbers are set in consultation with the Fire Authority.

Exceeding a maximum occupancy number will breach the licence conditions or Fire Certificate and could lead to a prosecution or revocation of the licence.

Without a valid licence, premises used for public entertainment will not be able to trade.

Occupancy levels are calculated on the number of people who can occupy a square metre of space; an individual being assumed to need a specific area of space.

The 'occupant load factor' (m^2 per person) varies according to the use of the premises.

Furniture and fixed seating are taken into account and in all-seater conference halls, for instance, maximum occupancy numbers are determined by the seating capacity.

Fire exit routes have to be of sufficient number and width to permit safe evacuation of all the permitted occupants (see table opposite).

Use of room or floor	Occupant load factor (m^2 per person)
Area for standing	0.3
Amusement arcade, assembly hall, bingo hall, club concourse, dance hall, pop concert venue, queuing area	0.5
Bar	0.3 to 0.5
Bowling alley, billiard room	9.3
Conference room, dining room, restaurant	1.0 to 1.5
Studio (film, TV, radio, recording)	1.4
Common room, reading room, staff room	1.0

Source: Home Office Guide 2000

Are there minimum widths of exit doors in premises used for public entertainment so that people can exit easily?

Yes, there are minimum widths for exit doors and escape routes so that the capacity of occupants can be evacuated without crushing.

In general terms and subject to Licensing or Fire Authority approval, the following provides a rule of thumb:

- minimum width of a single exit door should not be less than 750 mm
- each exit should be a minimum width of 1.05 m
- fire exit widths are calculated in unit widths of 525 mm.

In a crowded situation, a file of people moving through an opening will occupy 525 mm in width. This is known as the 'unit of exit width'. People will move through that space of 525 mm at a rate of 40 per minute.

The 'rate of discharge' from a premises is known as 40 persons per minute per unit of exit width.

When people are confronted with an exit door width of approximately 1.0 m they will pass through the door in two files, i.e. in an orderly two-by-two fashion. When exit door widths are wider than approximately 1.0 m people exit in a less orderly fashion and a homogeneous mass is formed.

What are maximum permissible evacuation times?

These are the times that it is assumed a maximum occupancy number for a building will take to evacuate. The time permitted will depend on the building classification and whether people are on the ground floor only or the ground and other floors.

Buildings which demonstrate a fire-resisting construction of at least 60 minutes are usually classified as Class A and the maximum permitted evacuation time is 3 *minutes*.

Buildings of traditional construction (e.g. brick, block, timber floors, timber roof trusses, etc.) are usually classified as Class B and the maximum permitted evacuation time is 2.5 *minutes*.

Temporary buildings, including marquees, and buildings with large quantities of wood, etc. would be classified as Class C and the maximum evacuation time would be 2.0 *minutes*.

Do fire precautions, maximum capacity numbers, etc. apply to smaller places of public entertainment such as village halls?

Yes. Fire Safety applies to all buildings in which people work or to which the public are admitted.

However, small premises may pose a slightly lower fire risk because numbers are less and there may be more management control of an event, etc.

Generally, if a 'small premises' use for public entertainment falls into the following categories, some relaxation of certain fire standards is allowed:

- the area of the hall, room or auditorium accommodation is 200 m^2 or less, and
- the premises is used by 300 members of the public or fewer, and
- no permanent provision is made for a closely seated audience.

Fire safety requirements for smaller premises will include:

- means of escape in case of fire
- fire-fighting equipment
- directional signs
- fire doors
- emergency lighting
- attendants
- procedures for calling the Fire Brigade
- use of flame-proof or flame-retardant furnishings.

Some permitted variations for small premises may be (subject to Fire Authority or Licensing Authority approval):

- provision of torches instead of emergency lighting for premises used occasionally for up to 100 people
- provision of battery and lamp units instead of emergency lighting as above
- provision of only one fire exit if the premises are on the ground floor and where there is adequate frontage and a final exit to a street, passageway or open space, the overall fire risk is low and the distance of travel to the final exit is not greater than 18 m, and curtains and furnishings, etc. comply with fire spread and retardancy requirements.

What precautions should be taken when hiring out premises to others for a public entertainment show, etc.?

Where premises are managed by voluntary committees, etc. and are hired out for functions it may be difficult for the licensee or manager to be present on the premises at every event.

It is recommended that when halls, etc. are hired out to organisations, the hirer fulfils the following conditions:

- is aged 18 years or over
- signs a written undertaking to accept responsibility for being in charge of and present on the premises at all times when the public are present and for ensuring that all conditions of the entertainment licence relating to management and supervision are met
- possesses a full copy of the licence conditions for the premises
- ensures that he receives instruction and training in the fire safety procedures and precautions for the building
- carries out a safety check of the building
- provides such attendants as may be required to assist in any emergencies while the public are on the premises.

What fire safety precautions are required in hotels and guest houses?

Hotels, guest houses, bed and breakfast establishments, hostels, etc. which have sleeping accommodation for *more than* six persons require a Fire Certificate.

Sleeping accommodation for staff is also included, so any staff quarters will be included in fire safety requirements.

If sleeping accommodation is provided above the first floor or below the ground floor for guests or staff, a Fire Certificate will be required.

Managers or owners of hotel accommodation should:

- ensure that adequate means of escape is available at all times
- train staff in the action to take in the event of a fire
- provide information to guests on the action to take in the event of a fire being discovered
- provide information to guests on what to do when the fire alarm sounds
- take adequate steps to prevent a fire from starting
- be aware of any special needs with regard to guests and staff, i.e. disabilities
- ensure that there is an adequate system for raising the alarm in the event of a fire
- ensure that suitable fire-fighting equipment is provided
- designate persons responsible for co-ordinating fire safety and emergency procedures on every shift.

Fire prevention procedures, fire safety management, training and maintenance and testing procedures which are applicable to all premises should be followed and practised in hotels and guest houses.

A Fire Risk Assessment must be carried out for every hotel, guest house, etc. and fire hazards and risks for both staff and guests must be identified. Suitable control measures must be implemented.

What areas of a hotel are most at risk from a fire?

Statistics show that most fires in hotels and guest houses occur in bedrooms and kitchens. Many fires do, however, start in bar areas, lounges and restaurants.

The misuse of cooking appliances is a common cause of fires in hotels.

The careless disposal of smoking materials is also another major cause of hotel fires.

A regular Fire Safety Check of the hotel should be carried out and fire prevention strategies implemented.

Routine maintenance programmes for all kitchen cooking and ventilation/extraction systems should be implemented, especially extractor hood or canopy and ductwork cleaning.

Furnishings, wall coverings, etc. should meet the appropriate British Standard for fire retardancy and spread of flame.

Fire retardant materials are not the same as fire resistant materials and the two should not be confused. Flame retardant means that the material will take some time to burn but will, nevertheless, at some stage, ignite and burn. Fire resistant materials will not burn.

What fire safety precautions apply to houses in multiple occupation?

Houses in multiple occupation (e.g. bedsits, shared student houses, common lodging houses, etc.), are considered to be very high risk in respect of fire.

The requirements for fire safety in houses in multiple occupation are covered in the Housing Act 1985 and other related Acts and Regulations and enforcement of fire safety falls to local authority Environmental Health Departments.

Houses in multiple occupation are categorised into different groups and fire precaution requirements will vary according to risk.

Generally, high risk houses in multiple occupation must:

- have a fire alarm
- have means of escape
- have protected staircases or alternative exit routes
- have smoke detectors or heat detectors
- have emergency exit signs
- have emergency lighting
- have an emergency plan

- have fire doors to all accommodation
- have cooking equipment in designated areas.

Local Authorities must consult with the Fire Authority for advice.

If a former hotel has been converted into a hostel for permanent residents it may have a Fire Certificate. This may still be enforced by the Fire Authority as the Fire Certificate would cover the new business although the Local Authority may have overall control with regard to legal compliance with fire safety legislation.

References

Fire Precautions Act 1971.

Fire Precautions (Hotels and Boarding Houses) Order 1972.

Fire Certificates (Special Premises) Regulations 1976.

Housing Act 1985: Houses in Multiple Occupation.

11

Fire safety training

What are an employer's responsibilities in respect of training employees in fire safety?

Employers have duties to provide employees with information, instruction and training in respect of all health and safety matters, including fire safety.

Section 2 of the Health and Safety at Work Etc. Act 1974 imposes a general duty on employers in respect of providing information, instruction and training.

Regulation 13 of the Management of Health and Safety at Work Regulations 1999 requires employers to provide training for employees as follows:

- upon recruitment and induction
- whenever an employee is exposed to new or increased risks from production, processes, plant or premises
- whenever training needs to be repeated or to be continuous so that the employee is up to date with current best practice.

Training must be flexible and adaptable to new working practices and consequential risks.

Training must be undertaken during working hours.

The Fire Precautions (Workplace) Regulations 1997 (amended) requires employers to provide training to any persons nominated to implement measures of fire safety.

Is it necessary to keep training records?

While the need for training records is not specifically laid down in law, it is good practice to keep records of who has received given training and when.

Employers will need to be able to demonstrate that they have complied with their legal duties and the best way to do so is to keep records.

If premises have a Fire Certificate there may be a requirement in the Schedule attached to the Certificate for fire safety training to be undertaken and records kept.

Are there any other circumstances under which an employer has a duty to train staff on fire safety?

Yes. If an owner or occupier of a building has submitted an application for a Fire Certificate, there is a duty under Section 5(A) of the Fire Precautions Act 1971 to ensure that interim fire safety measures are carried out.

Included in those interim measures is a specific duty to provide:

- to any persons employed to work in the premises instruction and training in what to do in case of a fire.

How often should fire safety training be carried out?

It is recommended that fire safety training is carried out at least twice a year.

The more thoroughly that staff are trained to be familiar with procedures for an emergency, the greater the chance that people will be evacuated safely and the threat to life safety reduced.

Fire safety training should always be carried out within the first day of a new employee starting work, or an existing employee being transferred to another part of the building.

Who can give training in fire safety?

The law expects 'competent' persons to be involved in giving advice and carrying out training in health and safety, including fire safety.

There are no specific qualifications needed in order to establish competency in delivering fire safety training but certain criteria should be demonstrated:

- experience in the workplace and the fire safety precautions in place
- knowledge of the fundamentals of fire and how and where it can start, its effects, etc.

It is not always necessary to give practical demonstrations in how to use fire extinguishers as practical videos can be used.

Is there a difference between induction training for fire safety and general fire safety training?

Induction training in fire safety is specified under the Management of Health and Safety at Work Regulations 1999 and generally is intended to cover key emergency information an employee needs to know in order to ensure their safety.

Induction fire safety should, at the very minimum, include:

- the method of raising the alarm
- evacuation procedures
- the location of means of escape
- the location of the assembly point
- checking in or clocking in procedure
- major fire hazards within the building.

Employees should be given induction training on the first day of their employment. It is good practice to develop everyone's basic knowledge by scheduling employees to attend more detailed fire safety training.

I intend to give my employees further fire safety training. What subjects should I cover?

Fire safety training need not be complicated. The following subjects should be included:

- the action to take on discovering a fire
- how to raise the alarm and what happens next
- the action to take on hearing the fire alarm
- the procedures for alerting customers and visitors including, where necessary, directing them to exits
- the arrangements for calling the Fire Brigade
- evacuation procedures for everyone to reach the assembly point at a safe place
- the location of the assembly point
- the location and, where appropriate, the use of fire-fighting equipment
- the location of the escape routes, especially those not in regular use
- how to open all escape doors including the use of any emergency fastenings, e.g. break glass bolts, etc.
- the importance of keeping fire doors closed to prevent the spread of fire, heat and smoke

- how to isolate machines, equipment, etc. as necessary
- how to isolate power supply sources
- why not to use lifts (if provided)
- evacuation of disabled customers or staff
- importance of general fire safety and good housekeeping
- understanding the Fire Risk Assessment, where it is kept, when it needs to be updated, etc.
- the control measures in place in the premises
- the need to report equipment faults and malfunctions
- how to deal with spillages.

Will Fire Wardens need additional fire safety training to that which they have had as an employee?

Probably. The topics for the Fire Warden (Marshall) fire safety course will be similar to those of any fire safety course already undertaken but the level of information that they will need may be greater.

Fire Wardens should be very aware of what their role is and must be clear about their responsibilities. So, the first aspect of Fire Warden training will be to state clearly that their role should be:

- to take appropriate and effective action if a fire occurs
- to ensure that escape routes are kept available for immediate use
- to identify and report fire hazards in the workplace.

The syllabus should then include:

- fire safety management within the company
  - the company Fire Safety Policy
  - the Fire Warden's role
  - developing a 'fire safe' environment
  - company fire procedures

- the nature of fire
    - the triangle of fire
- means of escape
    - fire warning
    - escape routes
    - emergency exits
    - signage
    - fire evacuation
    - fire drills
- fire prevention
    - good housekeeping
    - arson spotting
- fire fighting
    - location and type of fire extinguishers
    - how to use them
    - which fire extinguisher is used on each type of fire
    - fire blankets
    - use of sprinklers
    - how to identify fire extinguishers
- fire safety monitoring
    - fire safety checks
    - hazard spotting
    - fire prevention — electrical testing of appliances, etc.

What is the requirement for training in fire drills?

A practice fire drill should be carried out at least once every twelve months in most premises, and in those which are high risk or open to large numbers of the public (e.g. department stores), fire drills should be carried out more frequently.

Fire escape routes should all be available for practice fire drills, except that it is good practice to assume that one is inaccessible as would be likely in the event of a fire.

Even if members of the public are not present for the fire drill, it is sensible to carry one out because it is important that the staff know what to do in an emergency and that they are trained and familiar with the steps to be taken.

The full fire alarm should be activated by a member of staff as if it were an emergency and the procedures followed should be those that would be practised in a real emergency.

The results of a fire drill should be observed and monitored for efficiency and effectiveness. It is of little value if a fire drill is carried out which is ineffective and remedial measures are not taken to put it right.

It is important to check the time it takes to evacuate different parts of the building (e.g. floors), and the time it takes to evacuate the whole building. Is this in accordance with any times stipulated in the Fire Certificate? Is it reasonable? Does it leave people vulnerable?

Did people go to a place of relative safety? What happened to them once there? Were they assisted in the final exit?

Did everyone go to the assembly point?

Was anyone missing who should have been accounted for?

Did Fire Wardens check their floors or areas?

What happened to anyone with any disabilities? Could they have been effectively evacuated?

FIRE SAFETY STAFF INSTRUCTION AND TRAINING RECORD GENERAL RECORD

Name of employee	Date of training	Who gave the training?	What information was covered?	Signed (Employee)

EMPLOYEE RECORD FOR FIRE SAFETY TRAINING

Name of employee: ____________________

Job title: ____________________

Date of training: ____________________

Duration of training: ____________________

Training given by: ____________________

Subjects covered	Yes	No
Fire hazards		
Safe practices		
How to raise fire alarm		
Identifying the fire alarm sound		
Action to be taken on hearing alarm		
Calling the Fire Brigade		
Names and duties of Fire Wardens		
Location of fire-fighting equipment		
How to use equipment		
Other types of fire protection		
Escape routes to be used		
Fire doors — need to be kept shut		
Location of assembly point		
What not to do in a fire		

Subjects covered	**Yes**	**No**
Assisting people with disabilities		
Daily visual fire checks		
Reporting hazards, dangers, etc.		
Other: please specify		

Signed by Employee: ______________________________

Signed by Instructor: ______________________________

FIRE DRILL RECORD

Should be undertaken on a quarterly basis.

Date	Was all the building evacuated Yes/No	Did the drill include customers Yes/No	How long did it take to empty the building (minutes)	What do you need to do to improve anything for next time

Do employers have duties to train people other than their employees?

Employers must provide appropriate information, instruction and training to those who are affected by their business or undertaking.

Obviously, members of the public cannot be trained in fire safety but they can be provided with information in the shape of fire safety signs, etc.

Visitors to any business premises should be informed or instructed in fire safety procedures. This is often done at the Reception area and visitors are given name badges and brief information on what to do in the event of a fire. This can be on the reverse of the ID card or a separate instruction displayed on the counter.

Contractors working on the premises should be treated differently and they must be given information and instruction about the fire hazards and risks associated with the workplace. They must also be given information on emergency procedures, the fire alarm system, etc.

In addition, contractors must give information to the employer about any tasks that they are going to undertake which may cause hazards and risks to the employer's workforce.

Any use of Hot Works must be discussed and, where necessary, it would be good practice to run through a tool-box talk on the fire hazards, risks and control procedures for using Hot Works, including any Hot Works Permit procedure.

The overlap between employees working on the premises and contractors working on the premises in relation to health and safety is merging and it is now common practice for contractors and their employees and operatives to be invited to ongoing employer health and safety training courses.

As most fires in commercial buildings are caused during refurbishment works it makes good business sense to include contractors in fire safety training.

Top tips

- Keep training simple.
- Try to show a short fire safety video.
- Keep records.
- Train all new staff as soon as possible.
- Undertake refresher training.
- Train staff if anything changes in the premises, e.g. new extensions, etc.
- Make sure that staff know about the Fire Risk Assessments.
- Instruct or train contractors working on the premises.

Checklist for Fire Wardens

If a fire is discovered, the Fire Warden should:

- ensure that the fire alarm has been activated
- check that equipment and processes are shut down, where safe to do so
- evacuate people from the area for which they are responsible, or from the whole building, but without putting themselves at risk
- advise everyone to go to the assembly point
- assist any employees or visitors with disabilities and ensure that others help with their evacuation
- report to the 'Chief' Fire Warden that all are accounted for, or if anyone is missing
- report if any areas of their floor or building were not checked for people, i.e. any remote working areas
- guide employees back into the building when advised by the Fire Brigade that it is safe to do so.

Fire Wardens are *not* expected to fight the fire unless it is manageable and they feel competent to do so as a result of training.

References

Fire Precautions Act 1971.

Fire Precautions (Workplace) Regulations 1997 (amended).

Management of Health and Safety at Work Regulations 1999.

Health and Safety at Work Etc. Act 1974.

12

Fire safety for people with disabilities

As an employer, am I legally responsible for the safety of people with disabilities?

Yes. The law requires employers to be responsible for the health, safety and welfare of all employees and to make whatever special arrangements are necessary to deal with any employees who have 'special needs'.

Also, an employer has duties to persons not in his employ and must ensure that their safety is not put at risk by his undertaking.

Persons in control of premises must also consider the safety of persons 'resorting to the premises'.

Is there any other legislation applicable to people with disabilities?

The Disability Discrimination Act 1995 applies to all employers and service providers and requires them to treat people with disabilities 'no less favourably' than able-bodied persons.

Employers must not discriminate against people with disabilities and must ensure that adequate facilities are provided for them while they are at work. This will include procedures for fire safety, fire evacuation, etc.

Service providers include everyone who offers a service of some kind to the public or others and includes shops, banks, professional organisations, lawyers, insurance brokers, libraries, entertainment premises, offices, recreational buildings, etc.

Service providers must provide their service at no less a standard to people with disabilities than to non-disabled persons.

People with disabilities must have safe and reasonable access to goods and services. If there are physical barriers to access, then the service provider must make 'reasonable adjustments' to accommodate them. The law requires such adjustments to be made by October 2004.

What are disabilities?

Disabilities include physical impairment, lack of mobility, impaired hearing, impaired vision, mental illnesses and any condition which can be judged to have a long-term debilitating effect on the individual. The Code of Practice supporting the Disability Discrimination Act 1995 lists examples of disabilities. The list is not exhaustive and if any person believes that they have a disability and have been discriminated against, they can apply to the Disability Rights Commission for a judgement.

How do disabilities — both physical and other — affect fire safety requirements?

An employer and service provider must not discriminate against people with disabilities and this will also include in any emergency procedures.

Fire evacuation procedures, means of escape, etc. must address the needs of people with disabilities.

The Fire Risk Assessment required by the Fire Precautions (Workplace) Regulations 1997 (amended) must consider any additional fire safety risk to people with disabilities.

Control measures must be adopted which do not discriminate.

What special consideration needs to be given to people with disabilities in respect of fire safety?

The special considerations needed for people with disabilities are contained in the British Standard BS 5588: Part 8.

In summary, the following need to be considered in the Risk Assessment.

Means of escape

Door widths should be a minimum of 900 mm. Corridor widths need to be wide enough for wheelchair use with helper — approximately 1000 mm. Escape routes must be kept clear and properly signed. Any exit route especially designed for wheelchairs should be signed with a wheelchair symbol.

Escape routes should be on the same level — no steps. Ramps should be minimal and should not provide gradients steeper than 1 in 12.

Handrails should be provided along escape routes where practicable.

Disabled refuges

A refuge is a place of relative safety and is provided where people with disabilities may need to wait for further assistance, e.g. because they are unable to descend the stairs to the final exit.

A refuge must be in a fire protected area and separated from any fire by a fire resisting construction.

A refuge may be:

- a protected lobby
- a half landing on a staircase
- an open air area such as a flat roof, podium or similar which is sufficiently protected or remote from any fire risk and provided with its own means of escape.

It is advisable to consult with the Fire Authority on the location and use of disabled refuges.

People waiting in a disabled refuge must not cause obstructions to others exiting the premises.

A method of recording who is likely to be in the refuge must be in place so that the Fire Brigade can be advised immediately on their arrival at the emergency.

Fire alarms

People with impaired hearing may not be able to hear fire alarms and consideration must be given to providing flashing lights or visual fire alarms.

Again, the Fire Risk Assessment should consider the different types of alarms. Sometimes, high pitched fire alarms are audible to people with impaired hearing.

Fire signage

People with impaired sight may not be able to find their way out as they cannot see signs.

All signs must be in large lettering and easily visible from all parts of the building they serve.

Good lighting is essential and emergency lighting or photo-luminescent signs may assist a person deciphering instructions.

Colour contrasts may help people with impaired vision.

Handrails, steps, etc.

Handrails should be 'tactile' and protrude beyond the end of the fire escape route or stairs (without causing obstructions, etc.).

Steps and nosings should be colour contrasted (e.g. white nosings), as these help to make edges clear.

Tactile flooring may be suitable in front of exit routes or on ramps, etc.

Case study

A hotel company was required by the Disability Rights Commission to compensate a guest who felt discriminated against during a fire evacuation as he did not receive any specific information or assistance in evacuating his hotel room and was left unattended in what could have been a burning hotel. The hotel had to revise its procedures.

Fire evacuation notices

Instructions on how to respond in the event of a fire must be available for people with disabilities. It may be possible to give people with impaired vision Braille Fire Alarm Notices, or those with impaired hearing vibrating alarms.

Audio tapes may provide verbal instructions.

The use of pictogram signs help people to see easily which emergency route to take, etc.

The Emergency Plan

The Emergency Plan should include details of what procedures are to be followed in the event of an emergency and specific details of actions to be taken for people with disabilities must be included.

The Plan should include details of what staff should do when assisting in the evacuation of people with disabilities, e.g. a sighted person leading the visually impaired, etc.

Consideration should be given to how guide dogs for the blind and hard of hearing will be dealt with.

Lifts should not be used unless designed specifically for use in a fire, i.e. fire-fighting lifts.

Premises providing residential accommodation must have detailed Emergency Plans for evacuating people with disabilities.

References

Fire Precautions Act 1971.

Fire Precautions (Workplace) Regulations 1997 (amended).

British Standard (BS) 5588: Part 8: 1999.

Disability Discrimination Act 1995.

Fire safety — an employer's guide: HSE Books, Home Office.

Alphabetical list of questions